Algebra
SIMPLIFIED AND SELF-TAUGHT

David Frieder, M.A.
Adjunct Lecturer
New York University

Instructor of Mathematics
LaGuardia Community College

Director,
MATHWORKS
New York City

ARCO PUBLISHING, INC.
NEW YORK

To my parents, Philip and Henriette Frieder,
my sister, Eddie Frieder,
and my cat, Crayon

Published by Arco Publishing, Inc.
215 Park Avenue South, New York, N.Y. 10003

Copyright © 1983 by David Frieder

All rights reserved. No part of this book may
be reproduced, by any means, without permission
in writing from the publisher, except by a
reviewer who wishes to quote brief excerpts in
connection with a review in a magazine or
newspaper.

Library of Congress Cataloging in Publication Data

Frieder, David.
   Algebra simplified and self-taught.

  1. Algebra. I. Title.
QA152.2.F74 1983   512   83-2843
ISBN 0-668-05797-1 (pbk.)

Printed in the United States of America

CONTENTS

PREFACE v

ACKNOWLEDGMENTS vii

UNIT I. ALGEBRA 1

1. NUMERICAL EXPRESSIONS 2

1.1 Numerical Expressions, 2   1.2 Exponential Notation, 2
1.3 Square Roots, 3   1.4 Order of Operations, 5

2. ALGEBRAIC EXPRESSIONS 7

2.1 Algebraic Expressions, 7   2.2 Polynomials, 7   2.3 Evaluating Algebraic Expressions, 8   2.4 Evaluating the Subject of a Formula, 9

3. SIGNED NUMBERS 11

3.1 Signed Numbers, 11   3.2 Order Relationships on the Number Line, 12
3.3 Adding Signed Numbers, 12   3.4 Subtracting Signed Numbers, 14
3.5 Multiplying and Dividing Signed Numbers, 15   3.6 Evaluating Algebraic Expressions with Signed Numbers, 18

4. OPERATIONS WITH POLYNOMIALS 19

4.1 Adding and Subtracting Algebraic Terms (Monomials), 19
4.2 Adding and Subtracting Polynomials, 19   4.3 Multiplying and Dividing Monomials, 20   4.4 Multiplying Powers of the Same Base, 21
4.5 Dividing Powers of the Same Base, 22   4.6 Raising a Monomial to a Power, 23   4.7 Multiplying Polynomials by Monomials, 24
4.8 Dividing Polynomials by Monomials, 25   4.9 Multiplying Polynomials by Polynomials, 26

5. SOLVING LINEAR EQUATIONS AND INEQUALITIES 28

5.1 Solving Equations, 28   5.2 Solving Linear Equations in One Variable, 29   5.3 Solving Equations Containing Parentheses, 32
5.4 Solving Equations Containing Like Variable Terms, 33   5.5 Solving Equations Containing Fractions, 34   5.6 Solving Proportions, 36
5.7 Solving Equations Containing Square Roots, 37   5.8 Solving Literal Equations, 39   5.9 Solving Equations Simultaneously—Addition Method, 41   5.10 Solving Equations Simultaneously—Substitution Method, 44   5.11 Solving Linear Inequalities in One Variable, 46

6. FACTORING POLYNOMIALS 49

6.1 Factoring, 49   6.2 Factoring Polynomials Having Common Monomial Factor, 49   6.3 Factoring Trinomials of the General Form $ax^2 + bx + c$, 51   6.4 Factoring the Difference of Two Squares $A^2 - B^2$, 55   6.5 Factoring Expressions Completely, 57   6.6 Solving Quadratic Equations by Factoring, 58   6.7 Solving Incomplete Quadratic Equations, 63

7. OPERATIONS WITH SQUARE ROOTS   65

   7.1 Simplifying Square Roots of Numbers, 65   7.2 Simplifying Square Roots of Variables, 66   7.3 Adding and Subtracting Square Roots, 67   7.4 Multiplying Square Roots, 69   7.5 Dividing Square Roots, 70

8. ALGEBRAIC FRACTIONS   72

   8.1 Algebraic Fractions, 72   8.2 Reducing Algebraic Fractions to Lowest Terms, 72   8.3 Adding and Subtracting Like Algebraic Fractions, 74   8.4 Adding and Subtracting Unlike Algebraic Fractions, 75   8.5 Multiplying Algebraic Fractions, 78   8.6 Dividing Algebraic Fractions, 79   8.7 Simplifying Complex Algebraic Fractions, 80

ALGEBRA PRACTICE TEST   82

UNIT II. WORD PROBLEMS   88

9. ARITHMETIC PROBLEMS   89

   9.1 Fraction Word Problems, 89   9.2 Proportion Problems, 92   9.3 Percent Word Problems—General, 94   9.4 Percent Word Problems—Percent of Increase or Decrease, 97

10 NUMBER PROBLEMS   101

   10.1 Number Problems Involving One Unknown, 101   10.2 Number Problems Involving More Than One Unknown, 103   10.3 Consecutive Integer Problems, 104   10.4 Age Problems, 106

11. AVERAGE PROBLEMS   109

   11.1 Average Problems—Simple Average, 109   11.2 Average Problems—Weighted Average, 111   11.3 Mixture Problems, 112

12. MOTION PROBLEMS   115

   12.1 Motion Problems—Basic Formula, 115   12.2 Motion Problems—Special Situations, 118

13. WORK PROBLEMS   123

   13.1 Work Problems—Individuals, 123   13.2 Work Problems—Groups, 126

14. SET PROBLEMS   128

   14.1 Set Problems, 128

WORD PROBLEMS PRACTICE TEST   131

PREFACE

The purpose of this book is to provide the reader with a thorough review of elementary algebra and to show its applications to a wide variety of word problems found on various exams, including job application exams, the High School Equivalency Exam (GED), the Scholastic Aptitude Test (SAT), the Graduate Management Admission Test (GMAT), and the Graduate Record Examination General (GRE).

Written as a self-teaching guide, each section contains a large number of sample problems which are designed to illustrate the principles and procedures discussed in the text. The solutions to these problems are accompanied by detailed step-by-step explanations of how to proceed. Following the sample problems are similar practice problems. The answers to the practice problems are placed in the margin of the book for your convenience. If you prefer to do all the problems before checking them, you can cover the answers with an index card. In addition to the practice problems, a review test of twenty-five questions is given at the end of each unit. The answers to these questions, along with a worked out solution for each, follow the test.

ACKNOWLEDGMENTS

I would like to thank my family and friends for all the encouragement and support they gave me during the writing of this book. I would also like to give special thanks to Billy Karp and Jill Israel for providing me with a quiet and comfortable place to write the algebra chapter, to my partner Mark Weinfeld for his help in preparing the practice tests, and to my friends Marta Barszczewski, Steve Brauch, Stu Davis, Nancy Duggan, Michele Satty Gage, the Humphreys, Dennis Mangano, Doug Nervik, Emmanuelle Ortoli, and Danny and Cindy Shapiro for just being there. Finally, I would like to express my special gratitude to Marybeth Lahr for all the exceptional work she did in helping me write and edit the final manuscript.

UNIT I. ALGEBRA

In this unit, we will review the procedures for

- evaluating numerical and algebraic expressions
- performing operations with signed numbers
- performing operations with polynomials
- solving linear equations and inequalities
- factoring polynomials and solving quadratic equations
- performing operations with square roots
- performing operations with algebraic fractions

The emphasis in each section will be on algebraic manipulations and techniques. The application of algebra to a variety of word problems is presented in the next unit.

1

1.1 NUMERICAL EXPRESSIONS

In algebra, as in arithmetic, numerical expressions are formed with the fundamental operations of addition, subtraction, multiplication, and division. The symbols for these operations are summarized in the table below. Notice that multiplication and division can be denoted in several different ways.

Operations in Numerical Expressions

Addition: $8+3$
Subtraction: $9-7$
Multiplication: $5 \times 4,\ 5 \cdot 4,\ (5)(4),\ 5(4)$
Division: $2\overline{)6},\ 6 \div 2,\ \dfrac{6}{2}$

1.2 EXPONENTIAL NOTATION

As you recall, the numbers used in multiplication are called factors. When the same factor is repeated more than once, a special shorthand, called **exponential notation**, can be used. In this notation, the repeated factor, called the **base**, is written only once. Above and to the right of the base is written another number, called the **exponent** or **power** of the base, which indicates how many times the base is repeated as a factor. For example,

$3 \times 7 \times 2 = 42$

factors

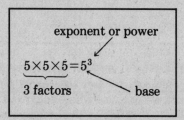

Products written in exponential notation are read by first naming the base and then the power. Some examples are given below. Note that the second power can also be read as "squared," and the third power can also be read as "cubed."

Product	Symbols	Words
8×8	8^2	8 to the second power, or 8 squared
$5 \times 5 \times 5$	5^3	5 to the third power, or 5 cubed
$7 \times 7 \times 7 \times 7$	7^4	7 to the fourth power

Since a product is always formed by at least two factors, the first power is somewhat ambiguous. Nevertheless, we define the first power to mean one "factor" of the base. For example, $6^1 = 6$, and $3^1 = 3$.

When evaluating products written in exponential notation, remember to write out as many factors of the base as indicated by the exponent, and then multiply those factors. A common mistake is to simply multiply the base by the exponent.

Correct: $6^2 = 6 \times 6 = 36$
Incorrect: $6^2 = 6 \times 2 = 12$

1.3 SQUARE ROOTS

The **square root** of a given number is a number whose square is equal to the given number. For example, the square root of 25, denoted, $\sqrt{25}$, is 5, because 5 squared is equal to 25. More examples of squares and square roots are given in the table below.

Squares	*Square Roots*
$1^2 = 1$	$\sqrt{1} = 1$
$2^2 = 4$	$\sqrt{4} = 2$
$3^2 = 9$	$\sqrt{9} = 3$
$4^2 = 16$	$\sqrt{16} = 4$
$5^2 = 25$	$\sqrt{25} = 5$
etc.	etc.

$\sqrt{289}$

$15^2 = 225$
$16^2 = 256$
$\boxed{17^2 = 289}$
$18^2 = 324$

Finding the square root of a number usually involves some form of trial and error. For example, to find the square root of 289 we keep squaring numbers until we find one whose square equals 289. As shown opposite, we find that $17^2 = 289$ and thus conclude that $\sqrt{289} = 17$.

Numbers such as 1, 4, 9, and 16, whose square roots are whole numbers, are called **perfect squares**. In the sample problem that follows, a procedure is demonstrated for finding the square roots of perfect squares. This procedure is based on the fact that if a number

M divides exactly M times into a number N, then M is the square root of N. For example, since 7 divides exactly 7 times into 49, then 7 is the square root of 49.

$$7\overline{)49}^{7} \quad \text{means} \quad \sqrt{49}=7$$

SAMPLE PROBLEM

Problem 1: Find $\sqrt{729}$

PROCEDURE	SOLUTION
(1) Choose any number, say 20, as a first estimate. Divide this estimate into 729.	(1) $\overline{}^{36}$ $20\overline{)729}$ $\underline{60}$ 129 $\underline{120}$ 9
(2) Disregarding the remainder, 9, find the average of the estimate, 20, and the quotient, 36.	(2) Average $=\dfrac{20+36}{2}$ $=28$
(3) Using the average as a new estimate, repeat the process until the estimate is the same as the quotient (until the remainder is 0).	(3) $28\overline{)729}^{26}$ $\underline{56}$ 169 $\underline{168}$ 1 $\text{Average}=\dfrac{28+26}{2}$ $=27$ $27\overline{)729}^{27}$ $\underline{54}$ 189 $\underline{189}$ 0 $\sqrt{729}=27 \qquad \textit{Answer}$

The procedures for finding the square roots of numbers which are not perfect squares are much more complicated and can only provide us with approximations such as $\sqrt{2}\approx 1.414$ and $\sqrt{3}\approx 1.732$. These procedures will not be shown here.

PRACTICE PROBLEMS

Answers			
(1) 19		(2) 21	
(3) 36		(4) 47	

Using the method of averaging, find the square root of each of the following perfect square numbers.

(1) $\sqrt{361}$ **(2)** $\sqrt{441}$ **(3)** $\sqrt{1296}$ **(4)** $\sqrt{2209}$

1.4 THE ORDER OF OPERATIONS

When numerical expressions contain more than one operation, there may be some ambiguity about the order in which the operations should be performed. For example, the expression $3+4\times2$ could either mean to add $3+4$ first, and then multiply by 2, or it could mean to multiply 4×2 first, and then add 3. Each order gives a different result:

$$\begin{array}{c|c} 3+4\times2 & 3+4\times2 \\ \stackrel{?}{=} 7\times2 & \stackrel{?}{=} 3+8 \\ \stackrel{?}{=} 14 & \stackrel{?}{=} 11 \end{array}$$

To avoid this ambiguity, a procedure has been established which specifies the exact order in which the operations should be performed. This procedure, called the **order of operations**, is given below.

To Evaluate a Numerical Expression:

(1) Perform all operations within *parentheses*, under *square root* symbols, and above and below *fraction bars*.
(2) Evaluate all *powers* and *square roots*.
(3) Perform all *multiplications* and *divisions* in the order they appear in the expression from left to right.
(4) Perform all *additions* and *subtractions* in the order they appear in the expression from left to right.

When we apply this procedure to the above expression, $3+4\times2$, we see that we should multiply 4×2 first, and that the correct result is 11.

SAMPLE PROBLEMS

Problem 2: What is the value of $7+3(8-3)^2$?

	PROCEDURE		SOLUTION	
(1)	Perform the subtraction within the parentheses.	(1)	$7+3(8-3)^2$ $=7+3(5)^2$	
(2)	Evaluate the power.	(2)	$=7+3(25)$	
(3)	Multiply. (Remember that $3(25)$ means 3×25.)	(3)	$=7+75$	
(4)	Add.	(4)	$=82$	Answer

Problem 3: Evaluate the expression $3+2\sqrt{1+6(4)}$

	PROCEDURE		SOLUTION	
(1)	Perform the operations under the square root sign. Remember to multiply $6(4)$ first.	(1)	$3+2\sqrt{1+6(4)}$ $=3+2\sqrt{1+24}$ $=3+2\sqrt{25}$	
(2)	Evaluate the square root.	(2)	$=3+2(5)$	
(3)	Multiply.	(3)	$=3+10$	
(4)	Add.	(4)	$=13$	Answer

PRACTICE PROBLEMS

Evaluate the following numerical expressions.

(1) $4^3-3(7-4)^2$

(2) $(2^3+3)(6^2-5^2)$

(3) $\dfrac{2^2+2^3}{3+3^2}$

(4) $\dfrac{2+6\sqrt{3^2+4^2}}{16}$

	Answers		
(1)	37	(2)	121
(3)	1	(4)	2

2

2.1 ALGEBRAIC EXPRESSIONS

$x^3 - y^2 + 7$
variables

Mathematical expressions which contain letters or other symbols to represent numbers are called **algebraic expressions**. These symbols are called either **literal numbers, placeholders, unknowns**, or most often, **variables**.

The following table shows that operations in algebraic expressions are denoted in essentially the same way as they are in numerical expressions. The only significant difference is that in algebraic expressions multiplication can also be denoted by placing the variables side by side, as in xy. This cannot be done in numerical expressions since placing numbers side by side would denote a new number, not multiplication. For example, placing 8 and 3 side by side would denote the number 83, not 8 times 3.

Operations in Algebraic Expressions

Addition: $x + y$

Subtraction: $x - y$

Multiplication: $x \cdot y, (x)(y), x(y), xy$

Division: $y \overline{)x}, x \div y, \dfrac{x}{y}$

2.2 POLYNOMIALS

$3 + 5y^2 - \dfrac{7x}{y}$
terms

$5a^2bx^3$
numerical coefficient

Algebraic expressions are formed by adding and subtracting basic units called **terms**. A term can either be a number, a variable, or the product or quotient of such quantities. The numerical factor of a term is called its **numerical coefficient** and is customarilly written on the left of the term. If a numerical coefficient is not indicated, it is understood to be 1. For example, $x^2y = 1x^2y$.

Algebraic expressions which do not include division by a variable or variables under square root signs are also called **polynomials**. For example, $3x^2 + 7x - 5$ and $x^2 + \dfrac{xy}{3}$ are polynomials, while $6x^2 + \dfrac{4}{y}$ and $3x^3 - \sqrt{x}$ are not polynomials. Polynomials having exactly one term are called **monomials**, two terms, **binomials**, and three terms, **trinomials**.

	Polynomials	
Monomials	Binomials	Trinomials
8	$9x-4$	$x+y-3$
$3x$	x^2-y^2	x^2+4x-1
$4x^2y^3$	$4x+y^3$	$x^2-3xy+y^2$

In the remainder of this section we will follow the common practice of using the word "monomial" to refer to expressions containing exactly one term, and the word "polynomial" to refer to expressions containing more than one term.

2.3 EVALUATING ALGEBRAIC EXPRESSIONS

Whenever specific numerical values are given for the variables in an algebraic expression, we can determine the value of the expression by the following procedure:

To Evaluate an Algebraic Expression:

(1) Replace the variables by the given values.
(2) Evaluate the resulting numerical expression according to the order of operations.

Remember that when substituting numbers into a product, use either parentheses or dots to indicate the product.

SAMPLE PROBLEMS

Problem 4: If $a=2$, $b=3$, and $x=1$, what is the value of a^3+5b^2-abx?

PROCEDURE	SOLUTION
(1) Replace the variables by their given numerical values.	(1) a^3+5b^2-abx $=(2)^3+5(3)^2-(2)(3)(1)$
(2) Evaluate the resulting numerical expression according to the order of operations.	(2) $=8+5(9)-(2)(3)(1)$ $=8+45-6$ $=47$ *Answer*

Problem 5: Evaluate the expression $\dfrac{\sqrt{b^2-4ac}}{2a}$ for the values $a=6$, $b=11$, and $c=3$.

PROCEDURE	SOLUTION
(1) Replace the variables by their given numerical values.	(1) $\dfrac{\sqrt{b^2-4ac}}{2a}$ $=\dfrac{\sqrt{11^2-4\cdot 6\cdot 3}}{2\cdot 6}$
(2) Evaluate the resulting numerical expression according to the order of operations.	(2) $=\dfrac{\sqrt{121-72}}{12}$ $=\dfrac{\sqrt{49}}{12}$ $=\dfrac{7}{12}$ *Answer*

PRACTICE PROBLEMS

For $a=2$, $b=3$, and $x=1$, evaluate the following expressions.

(1) $7a^3bx^2$

(2) $b^3-5(a+x)$

(3) $\dfrac{6x^2+2b^2}{4a}$

(4) $\dfrac{\sqrt{b^2+a^4}}{a+bx}$

Answers
(1) 168   (2) 12
(3) 3   (4) 1

2.4 EVALUATING THE SUBJECT OF A FORMULA

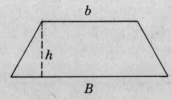

A **formula** is a mathematical relationship between one variable, called the **subject** of the formula, and one or more other variables. For example, the area of a trapezoid is related to its height and two bases by the formula

$$A=\left(\dfrac{B+b}{2}\right)h.$$

$A=\left(\dfrac{B+b}{2}\right)h$

$=\left(\dfrac{6+4}{2}\right)3$

$=\left(\dfrac{10}{2}\right)3$

$=15$

To determine the value of the subject of a formula, follow the same procedure used when evaluating an algebraic expression. Replace all the other variables in the formula by specific numerical values, and then evaluate the resulting numerical expression. For example, if the height of a trapezoid is 3, and the bases are 6 and 4, then, as shown opposite, the area is 15.

SAMPLE PROBLEM

Problem 6: For the values $b=7$ and $h=6$, find the volume of a pyramid given by the formula below.

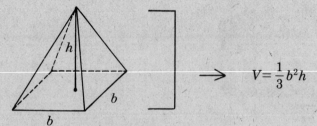

$$V = \frac{1}{3}b^2h$$

PROCEDURE		SOLUTION
(1) Replace b and h by their given numerical values, and evaluate the resulting expression.	(1)	$V = \frac{1}{3}b^2h$ $= \frac{1}{3}(7)^2(6)$ $= \frac{1}{3}(49)(6)$ $= 98$ Answer

3

3.1 SIGNED NUMBERS

Operations

Addition: $5+3$
Subtraction: $7-2$

Signs

Positive Sign: $+6$
Negative Sign: -4

The $+$ and $-$ symbols, which are used to represent the **operations** of addition and subtraction, can also be used as **signs** to indicate whether a number is greater than zero or less than zero. When used in this way, the $+$ and $-$ symbols are called **positive** and **negative** signs, and the resulting numbers are called **signed numbers**.

Signed numbers can be represented graphically on a scale called the **number line**, which is constructed in the following way. An arbitrary point is chosen to represent the number 0. This point is called the **origin**. Using a specified unit of measurement, points are marked off to the right and the left of the origin. Positive whole numbers are placed at the points to the right of the origin; negative whole numbers are placed at the points to the left of the origin. (Positive numbers can be written with or without the $+$ sign.) In the diagram below, arrows are drawn at the ends of the line to indicate that the line extends indefinitely in both directions.

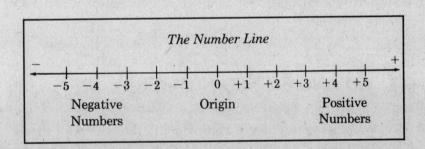

The Number Line

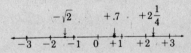

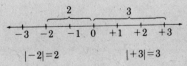

The set of numbers marked off on the number line, $\{\ldots, -3, -2, -1, 0, +1, +2, +3, \ldots\}$, is called the set of **integers**. Although only integers are indicated, all fractions and decimals, both positive and negative, are also on the line. Some examples are shown opposite.

The distance from any given number on the line to the origin is called the **absolute value** of the number. For example, since $+3$ is 3 units from the origin, its absolute value is 3. Similarly, since -2 is 2 units from the origin, its absolute value is 2. The absolute value of a number is denoted by placing the number between two vertical bars. We would write $|+3|=3$ and $|-2|=2$. Notice that the absolute value of a number is simply the number without its sign.

11

3.2 ORDER RELATIONSHIPS ON THE NUMBER LINE

The numbers are arranged on the number line in increasing order from left to right. Consequently, if a number x lies to the left of a number y, then x is *less than* y. This is written $x < y$. Equivalently, if a number y lies to the right of a number x, then y is *greater than* x. This is written $y > x$. In addition, if a number x lies to the left of a number y ($x < y$), and the number y lies to the left of a number z ($y < z$), then y lies *between* x and z. This is written $x < y < z$.

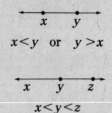

Numbers can be related in more than one way. For example, if a number x is either less than or equal to a number y, then we can combine the symbols $<$ and $=$ and write $x \leq y$. Similarly, if a number x is either greater than or equal to a number y, then we can combine the symbols $>$ and $=$ and write $x \geq y$.

The symbols for the different order relationships are summarized below.

Order Relationships	
Symbol	Words
$x = y$	x is equal to y
$x < y$	x is less than y
$x > y$	x is greater than y
$x \leq y$	x is less than or equal to y
$x \geq y$	x is greater than or equal to y

Any of these symbols can be negated simply by drawing a slash through it. For example, $x \neq y$ means x is *not* equal to y, and $x \not< y$ means x is *not* less than y.

3.3 ADDING SIGNED NUMBERS

The operation of addition is not performed the same with signed numbers as it is in ordinary arithmetic, where all the numbers are positive. Instead, "adding" is more like "combining," in the sense of combining profit and loss. In each of the examples below a positive number represents a profit, and a negative number represents a loss. The "sum" of the numbers is then the combined, or net, result.

$+8$ profit	-8 loss	$+8$ profit	-8 loss
$+6$ profit	-6 loss	-6 loss	$+6$ profit
$+14$ profit	-14 loss	$+2$ profit	-2 loss

In the first two examples, where the signs of the numbers are the same, the "sum" is obtained by adding the numbers. However, in the last two examples, where the signs of the numbers are different, the "sum" if obtained by subtracting the numbers. In general, we use the following procedure:

$(+9)+(+5)=+14$

$(-9)+(-5)=-14$

$(+9)+(-5)=+4$

$(-9)+(+5)=-4$

> **To Add (Combine) Two Signed Numbers:**
> (1) If the signs of the numbers are the *same*, add their absolute values (the numerical parts of the numbers, without their signs), and keep the common sign.
> (2) If the signs of the numbers are *different*, subtract the smaller absolute value from the larger absolute value, and keep the sign of the number with the larger absolute value.

$(+9)$ + (-5)
↗   ↑   ↖
positive add negative
sign   sign

Note that when we write signed number addition horizontally, we place the signed numbers in parentheses. This is done to distinguish the positive and negative signs of the numbers from the addition symbol. The same thing will be done in subtraction of signed numbers.

To add more than two signed numbers, proceed step by step, adding two at a time until reaching the final result:

$$(+6)+(-5)+(+3)+(-9)$$
$$=(+1)+(+3)+(-9)$$
$$=(+4)+(-9)$$
$$=-5$$

An alternate way of proceeding is to add the positive numbers and negative numbers separately, then combining the resulting sums:

$$(+6)+(-5)+(+3)+(-9)$$
$$=(+9)+(-14)$$
$$=-5$$

PRACTICE PROBLEMS

Answers	
(1) −5	(2) −13
(3) 0	(4) +5

Perform the following additions.

(1) $(+6)+(-4)+(-7)$

(2) $(-3)+(-8)+(-2)$

(3) $(-5)+(-4)+(+9)$

(4) $(+6)+(-2)+(+1)$

3.4 SUBTRACTING SIGNED NUMBERS

Subtracting signed numbers is analogous to dividing fractions in arithmetic. For example, when we divide $\frac{5}{7}$ by $\frac{3}{4}$, we don't actually divide the fractions, but instead, change the division to its inverse, multiplication, and change the divisor to its reciprocal, $\frac{4}{3}$. That is,

$$\frac{5}{7} \div \frac{3}{4}$$

$$= \frac{5}{7} \times \frac{4}{3}$$

$$= \frac{20}{21}$$

Similarly, when we subtract signed numbers, we don't actually subtract the numbers, but instead, change the subtraction to its inverse, addition, and change the subtrahend (the number being subtracted) to its opposite signed number.

$(+8)-(+6)$	$(-8)-(-6)$	$(+8)-(-6)$	$(-8)-(+6)$
$\downarrow \downarrow$	$\downarrow \downarrow$	$\downarrow \downarrow$	$\downarrow \downarrow$
$=(+8)+(-6)$	$=(-8)+(+6)$	$=(+8)+(+6)$	$=(-8)+(-6)$
$= +2$	$= -2$	$= +14$	$= -14$

To Subtract Signed Numbers:

(1) Change all subtractions to addition.
(2) Change the sign of each number being subtracted (each number following a subtraction symbol) to its opposite sign.
(3) Follow the procedure for adding (combining) signed numbers.

SAMPLE PROBLEM

Problem 7: Evaluate the expression $(+2)-(-4)+(-3)-(+1)$.

PROCEDURE

(1) Change all subtractions to addition, and change the sign of each number being subtracted to its opposite sign. Add (combine) the resulting signed numbers.

SOLUTION

(1) $(+2)-(-4)+(-3)-(+1)$
$\downarrow \downarrow \downarrow \downarrow$
$=(+2)+(+4)+(-3)+(-1)$
$= (+6)+(-3)+(-1)$
$= (+3)+(-1)$
$= +2$ *Answer*

PRACTICE PROBLEMS

Perform the following additions and subtractions.

(1) $(-6)-(+5)-(-2)-(-1)$

(2) $(+6)-(-2)-(+7)-(-4)$

(3) $(-8)-(-5)+(-1)-(+3)$

(4) $(4)+(-5)-(7)+(2)$

Answers			
(1)	−8	(2)	+5
(3)	−7	(4)	−6

3.5 MULTIPLYING AND DIVIDING SIGNED NUMBERS

Unlike addition and subtraction of signed numbers, multiplication and division are performed in almost the same manner as ordinary arithmetic.

Same Signs

$(+5)(+2) = +10$
$(-5)(-2) = +10$

$\dfrac{(+8)}{(+2)} = +4$

$\dfrac{(-8)}{(-2)} = +4$

Different Signs

$(+5)(-2) = -10$
$(-5)(+2) = -10$

$\dfrac{(+8)}{(-2)} = -4$

$\dfrac{(-8)}{(+4)} = -4$

To Multiply or Divide Two Signed Numbers:

(1) Multiply or divide the absolute values of the numbers (ignoring their signs).
(2) If the signs of the numbers are the *same*, make the sign of the result *positive*. That is,

$$(+)(+) \text{ or } (-)(-) = (+)$$

$$\frac{(+)}{(+)} \text{ or } \frac{(-)}{(-)} = (+)$$

If the signs of the numbers are *different*, make the sign of the result *negative*. That is,

$$(+)(-) \text{ or } (-)(+) = (-)$$

$$\frac{(+)}{(-)} \text{ or } \frac{(-)}{(+)} = (-)$$

SAMPLE PROBLEM

Problem 8: Evaluate the expression $\dfrac{(-8)}{(-2)} + (-3)(5) - \dfrac{(+6)}{(-2)}$.

PROCEDURE	SOLUTION
(1) Multiply and divide the signed numbers.	(1) $\dfrac{(-8)}{(-2)} + (-3)(5) - \dfrac{(+6)}{(-2)}$ $= (+4) + (-15) - (-3)$
(2) Change subtraction to addition of the opposite signed number, and add (combine) the resulting numbers.	(2) $= (+4) + (-15) + (+3)$ $= (-11) + (+3)$ $= -8$ *Answer*

To multiply more than two signed numbers, we proceed step by step, multiplying two numbers at a time, until we reach the final result:

$$(-3)(+5)(-4)(-2)$$
$$= (-15)(-4)(-2)$$
$$= (+60)(-2)$$
$$= -120$$

An alternate way of determining the final sign of a product of two or more signed numbers is based on the number of negative factors in the original expression. Since each pair of negative factors gives a positive result, the only way that the final sign can be negative is if the original expression contains an odd number of negative factors. Therefore,

---

To Determine the Final Sign of a Product of Signed Numbers:

(1) Count the number of *negative factors*.
(2) If this number is *even*, make the sign of the final result *positive*.
If this number is *odd*, make the sign of the final result *negative*.

Two Negative Factors
$(-2)(+5)(-3) = +30$

Three Negative Factors
$(-2)(-5)(-3) = -30$

This procedure is particularly useful for determining the final sign of a signed number raised to a power:

$$(+2)^4 = \underbrace{(+2)(+2)(+2)(+2)}_{\text{no negative factors}} = +16$$

$$(+2)^5 = \underbrace{(+2)(+2)(+2)(+2)(+2)}_{\text{no negative factors}} = +32$$

$$(-2)^4 = \underbrace{(-2)(-2)(-2)(-2)}_{\text{4 negative factors}} = +16$$

$$(-2)^5 = \underbrace{(-2)(-2)(-2)(-2)(-2)}_{\text{5 negative factors}} = -32$$

The only combination of base and power that results in a negative final sign is a *negative base* raised to an *odd power*. All other combinations result in a positive final sign.

Powers of Signed Numbers	
Positive Base	Negative Base
$(+)^{\text{even}} = (+)$	$(-)^{\text{even}} = (+)$
$(+)^{\text{odd}} = (+)$	$(-)^{\text{odd}} = (-)$

$(-5)^2 = (-5)(-5)$
$\quad\quad = +25$

$-5^2 = -(5)(5)$
$\quad\quad = -25$

Note that expressions like $(-5)^2$ and -5^2 do *not* mean the same thing. The first, $(-5)^2$, means to square the number -5. This gives $+25$. The second, -5^2, means the negative of the number 5^2. This gives -25.

PRACTICE PROBLEMS

Evaluate the following expressions.

Answers			
(1)	-168	(5)	-27
(2)	-5	(6)	$+1$
(3)	-14	(7)	$+64$
(4)	44	(8)	-64

(1) $(-3)(+4)(-2)(-7)$

(2) $\dfrac{(-6)(+5)}{(-3)(-2)}$

(3) $\left(\dfrac{-16}{+2}\right) + (-3)(+2)$

(4) $(-5)(-2)(3) - (7)(-2)$

Evaluate the following powers.

(5) $(-3)^3$ (6) $(-1)^{18}$ (7) $(-2)^6$ (8) -2^6

3.6 EVALUATING ALGEBRAIC EXPRESSIONS WITH SIGNED NUMBERS

Remember, when evaluating an algebraic expression for specific values of the variables, substitute the values into the expression and then follow the order of operations. When substituting signed numbers into algebraic expressions, it is advisable to use parentheses to distinguish the positive and negative signs from the operations.

SAMPLE PROBLEMS

Problem 9: For $a=3$, $b=-2$, and $x=-4$, find the value of $b^3 - ax + x^2$.

PROCEDURE

(1) Substitute the given values into the expression.

(2) Evaluate the powers and the product.

(3) Change subtraction to addition of the opposite signed number, and add (combine) the resulting numbers.

SOLUTION

(1) $b^3 - ax + x^2$
$= (-2)^3 - (3)(-4) + (-4)^2$

(2) $= (-8) - (-12) + (+16)$
$\qquad\downarrow\quad\downarrow$

(3) $= (-8) + (+12) + (+16)$
$= \quad (+4) + (+16)$
$= \quad +20 \qquad$ Answer

Problem 10: For $x=3$ and $y=-2$, evaluate $x^2 - 5y$.

PROCEDURE

(1) Substitute the given values into the expression.

(2) Evaluate the power and product.

(3) Change subtraction to addition of the opposite signed number, and add the resulting numbers. (Note that instead of subtracting $5(-2)$, we could multiply $-5(-2)$. This would give $9+10$, or 19, the same result as before.)

SOLUTION

(1) $x^2 - 5y$
$= (3)^2 - 5(-2)$

(2) $= (9) - (-10)$
$\qquad\downarrow\quad\downarrow$

(3) $= (9) + (+10)$
$= +19 \qquad$ Answer

PRACTICE PROBLEMS

For $a=-2$, $b=-5$, and $x=3$, evaluate the following expressions.

(1) $3a^3 - b$

(2) $a^2 + b^2 - x^2$

(3) $\dfrac{abx}{a+b-x}$

(4) $-b^2 - 4a$

Answers
(1) -19 (2) 20
(3) -3 (4) -17

4

4.1 ADDING AND SUBTRACTING ALGEBRAIC TERMS (MONOMIALS)

As you recall, algebraic terms are formed by the product and quotient of numbers and variables. Terms which have identical variable factors are called **like terms**. Some examples are given below.

Like Terms	Unlike Terms
$5x$ and $-7x$	$7x$ and $3y$
$-2x^2y$ and $3x^2y$	$-2ax$ and $3bx$
a^2b^3 and $-\frac{1}{4}a^2b^3$	$3x^2$ and $5x^3$

To add and subtract like terms, simply add and subtract (combine) their numerical coefficients. The result is a single term having the same variable factors as the original like terms.

$$3x^2y + 2x^2y - 9x^2y$$
$$= (3 + 2 - 9)x^2y$$
$$= -4x^2y$$

Terms which do not have identical variable factors (unlike terms) cannot be combined into a single term. For example, expressions such as $3x + 4y$ and $7xy^2 - 5xy$ cannot be simplified further.

PRACTICE PROBLEMS

Perform the following additions and subtractions.

(1) $6x^3 - 2x^3 - 7x^3 + x^3$
(2) $4xy^2 - 3x^2y + xy^2 - 4x^2y$
(3) $(5x) + (-8x) - (-3x) + (x)$
(4) $(3x) + (-5y) - (x) - (-7y)$

Answers
(1) $-2x^3$
(2) $5xy^2 - 7x^2y$
(3) x
(4) $2x + 2y$

4.2 ADDING AND SUBTRACTING POLYNOMIALS

When adding and subtracting algebraic expressions containing more than one term (polynomials), it is usually helpful to rearrange the expressions in a vertical format with the like terms lined up. Remember that to subtract, first change the sign of every term being subtracted to its opposite sign.

SAMPLE PROBLEMS

Problem 11: Add $(3x^2-4x-7)+(x^2+6x-2)$.

PROCEDURE	SOLUTION
(1) Rearrange the expressions vertically, with the like terms lined up. Combine the numerical coefficients of the like terms.	(1) $\quad 3x^2-4x-7$ $\quad\quad\; x^2+6x-2$ $\quad\quad\overline{4x^2+2x-9}\quad$ Answer

---

Problem 12: Subtract $(4x^2-5xy-y^2)-(x^2+3xy-4y^2)$.

PROCEDURE	SOLUTION
(1) Change subtraction to addition, and change the sign of each term being subtracted to its opposite sign.	(1) $(4x^2-5xy-y^2)-(x^2+3xy-4y^2)$ $\quad=(4x^2-5xy-y^2)+(-x^2-3xy+4y^2)$
(2) Line up the like terms, and combine the numerical coefficients.	(2) $\quad 4x^2-5xy-y^2$ $\quad\quad -x^2-3xy+4y^2$ $\quad\quad\overline{3x^2-8xy+3y^2}\quad$ Answer

---

PRACTICE PROBLEMS

Perform the following additions and subtractions.

(1) $(5x^2+6x-3)+(2x^2-8x-5)$
(2) $(y^2-2y+8)-(3y^2-5y-1)$
(3) $(x^2-3xy+5y^2)+(5x^2-3y^2)$
(4) $(3x^2+5xy)-(4x^2-2y^2)$

> **Answers**
> (1) $7x^2-2x-8$
> (2) $-2y^2+3y+9$
> (3) $6x^2-3xy+2y^2$
> (4) $-x^2+5xy+2y^2$

4.3 MULTIPLYING AND DIVIDING MONOMIALS

Although addition and subtraction can only be performed with like terms, multiplication and division can be performed with any type of terms, like or unlike. We simply regroup the numerical and variable factors, and then multiply or divide them separately:

Multiplication	Division
$(-5x^2)(2y^3) = (-5)(2)(x^2)(y^3)$	$\dfrac{-6x^3}{2y} = \left(\dfrac{-6}{2}\right)\left(\dfrac{x^3}{y}\right)$
$= -10x^2 y^3$	$= \dfrac{-3x^3}{y}$

4.4 MULTIPLYING POWERS OF THE SAME BASE

When multiplying monomials which contain the same variable factor, we can simplify the result in the following way:

$$(3x^2)(2x^3)(5x^2) = (3)(2)(5)(x^2)(x^3)(x^2)$$
$$= 30\,\overbrace{(x \cdot x)}^{2}\overbrace{(x \cdot x \cdot x)}^{3}\overbrace{(x \cdot x)}^{2}$$
$$= 30\,\overbrace{(x \cdot x \cdot x \cdot x \cdot x \cdot x \cdot x)}^{7 \text{ factors}}$$
$$= 30x^7$$

By writing out all the variable factors in the product, we see that the exponent in the result is simply the *sum* of the individual exponents.

In general, when multiplying m factors of x by n factors of x, the result will contain a total of $m+n$ factors of x. In other words,

$x^2 \cdot x^3 = x^{2+3} = x^5$

$y^2 \cdot y^5 \cdot y = y^{2+5+1} = y^8$

> *To Multiply Powers of the Same Base*:
>
> Keep the base and *add* the exponents.
>
> $$x^m \cdot x^n = x^{m+n}$$

SAMPLE PROBLEM

Problem 13: Multiply $(-3x^2 y)(2x^3 y^2)(5xy)$.

PROCEDURE	SOLUTION
(1) Multiply the numerical factors and variable factors separately.	(1) $(-3x^2 y)(2x^3 y^2)(5xy)$ $= (-3)(2)(5)(x^2 \cdot x^3 \cdot x)(y \cdot y^2 \cdot y)$
(2) For the variables with the same base, add the exponents. Remember that when an exponent does not appear, it is understood to be 1.	(2) $= -30(x^{2+3+1})(y^{1+2+1})$ $= -30x^6 y^4$ Answer

PRACTICE PROBLEMS

Perform the following multiplications.

(1) $(-2x^2)(5x^3)$
(2) $(-7x^3y^2)(-6xy)$
(3) $(-3xy)(5xy^2)(2x)$
(4) $(6a^2b)(ax)(2bx^2)$

Answers	
(1)	$-10x^5$
(2)	$42x^4y^3$
(3)	$-30x^3y^3$
(4)	$12a^3b^2x^3$

4.5 DIVIDING POWERS OF THE SAME BASE

The procedure for dividing powers of the same base is similar to the one for multiplying powers of the same base. However, instead of adding the exponents, subtract them.

Higher Power in Numerator

$$\frac{18x^5}{3x^2} = \left(\frac{18}{3}\right)\left(\frac{x^5}{x^2}\right)$$

$$= 6\left(\frac{x \cdot x \cdot x \cdot \cancel{x} \cdot \cancel{x}}{\cancel{x} \cdot \cancel{x}}\right)$$

$$= \frac{6x^3}{1}$$

$$= 6x^3$$

Higher Power in Denominator

$$\frac{4x^2}{8x^6} = \left(\frac{4}{8}\right)\left(\frac{x^2}{x^6}\right)$$

$$= \left(\frac{1}{2}\right)\left(\frac{\cancel{x} \cdot \cancel{x}}{x \cdot x \cdot x \cdot x \cdot \cancel{x} \cdot \cancel{x}}\right)$$

$$= \left(\frac{1}{2}\right)\left(\frac{1}{x^4}\right)$$

$$= \frac{1}{2x^4}$$

This time, notice that some of the factors cancel, resulting in an exponent which is the *difference* of the individual exponents. Also notice that the variable in the result is either in the numerator or denominator, depending upon which of the original factors has the higher exponent (more factors). In general,

To Divide Powers of the Same Base:

Keep the base and *subtract* the exponents.

$$\frac{x^m}{x^n} = \begin{cases} x^{m-n}, & \text{for } m > n \\ 1, & \text{for } m = n \\ \frac{1}{x^{n-m}}, & \text{for } m < n \end{cases}$$

$$\frac{x^5}{x^2} = x^{5-2} = x^3$$

$$\frac{x^4}{x^4} = 1$$

$$\frac{x^3}{x^7} = \frac{1}{x^{7-3}} = \frac{1}{x^4}$$

SAMPLE PROBLEM

Problem 14: Divide $\dfrac{6x^2y^6z}{8x^5y^2z^2}$.

PROCEDURE	SOLUTION
(1) Separate the numerical and variable factors.	(1) $\dfrac{6x^2y^6z}{8x^5y^2z^2} = \left(\dfrac{6}{8}\right)\left(\dfrac{x^2 \cdot y^6 \cdot z}{x^5 \cdot y^2 \cdot z^2}\right)$
(2) Reduce the numerical fraction. For each variable factor with the same base, subtract the smaller exponent from the larger exponent.	(2) $= \left(\dfrac{\cancel{6}^{\,3}}{\cancel{8}_{\,4}}\right)\left(\dfrac{1 \cdot y^{6-2} \cdot 1}{x^{5-2} \cdot 1 \cdot z^{2-1}}\right)$
	$= \dfrac{3y^4}{4x^3z}$ Answer

PRACTICE PROBLEMS

Perform the following divisions.

(1) $\dfrac{-16x^3y}{10xy}$

(2) $\dfrac{2x^2y^3}{6x^3y^5}$

(3) $\dfrac{6xy^3}{4x^2y}$

(4) $\dfrac{-2a^3b^5x}{8ab^2x^3}$

Answers

(1) $\dfrac{-8x^2}{5}$

(2) $\dfrac{1}{3xy^2}$

(3) $\dfrac{3y^2}{2x}$

(4) $\dfrac{-a^2b^3}{4x^2}$

4.6 RAISING A MONOMIAL TO A POWER

Raising a monomial to a power is a special case of multiplying monomials. In the example below, the exponent outside the parentheses indicates that we should multiply 3 factors of the monomial inside the parentheses.

$$(x^4y^2)^3 = (x^4y^2)(x^4y^2)(x^4y^2)$$
$$= (x^4 \cdot x^4 \cdot x^4)(y^2 \cdot y^2 \cdot y^2)$$
$$= (x^{\overbrace{4+4+4}^{4\cdot 3}})(y^{\overbrace{2+2+2}^{2\cdot 3}})$$
$$= x^{12}y^6$$

The exponent of each factor in the result is the product of the exponent inside the parentheses and the exponent outside the parentheses. In general,

> *To Raise a Power of a Given Base to Another Power*:
> Keep the base and *multiply* the exponents.
> $(x^m)^n = x^{m \cdot n}$

$$(3x^2y^5)^4 = 3^{1 \cdot 4} x^{2 \cdot 4} y^{5 \cdot 4}$$
$$= 3^4 x^8 y^{20}$$
$$= 81 x^8 y^{20}$$

PRACTICE PROBLEMS

Raise each of the following monomials to the power indicated.

(1) $(3x^2y^5)^2$
(2) $(-2xy^2)^3$
(3) $(-x^2y^3)^2$
(4) $(5x^2y^7)^3$

> Answers
> (1) $9x^4y^{10}$
> (2) $-8x^3y^6$
> (3) x^4y^6
> (4) $125x^6y^{21}$

4.7 MULTIPLYING POLYNOMIALS BY MONOMIALS

To multiply a polynomial by a monomial, we use the **distributive law of multiplication**.

> *Distributive Law of Multiplication*
> When multiplying a sum of terms by a single term, multiply each term in the sum by the single term, and then add the results.
> $A(B + C + \cdots + R) = AB + AC + \cdots + AR$

$3a(4x + 2y)$
$= 3a(4x) + 3a(2y)$
$= 12ax + 6ay$

SAMPLE PROBLEM

Problem 15: Multiply $3x^2y(5x^3y^2 - 2xy)$.

PROCEDURE	SOLUTION
(1) Using the distributive law, multiply each term in the parentheses by $3x^2y$. Remember that for variables with the same base, add the exponents.	(1) $3x^2y(5x^3y^2 - 2xy)$ $= 3x^2y(5x^3y^2) + 3x^2y(-2xy)$ $= 15x^5y^3 - 6x^3y^2$ Answer

Operations with Polynomials / 25

PRACTICE PROBLEMS

Perform the following multiplications.

Answers
(1) $6x^3 - 10xy$
(2) $-6x^2y - 9xy^2$
(3) $-2x^3y^2 + 8xy^5$
(4) $18a^3b^3x + 12a^3b^4x$

(1) $2x(3x^2 - 5y)$
(2) $-3xy(2x + 3y)$
(3) $-2xy^2(x^2 - 4y^3)$
(4) $6a^2b^3(3ax + 2bx)$

4.8 DIVIDING POLYNOMIALS BY MONOMIALS

To divide a polynomial by a monomial, we use the **distributive law of division**.

$$\frac{8x + 6y}{2a}$$
$$= \frac{8x}{2a} + \frac{6y}{2a}$$
$$= \frac{4x}{a} + \frac{3y}{a}$$

Distributive Law of Division

When dividing a sum of terms by a single term, divide each term in the sum by the single term, and then add the results.

$$\frac{B + C + \cdots + R}{A} = \frac{B}{A} + \frac{C}{A} + \cdots + \frac{R}{A}$$

SAMPLE PROBLEM

Problem 16: Divide $\dfrac{6xy^4 + 15x^3y}{3x^2y^2}$

PROCEDURE

(1) Using the distributive law, divide each term in the numerator by $3x^2y^2$. Remember that for variables with the same base, subtract the exponents.

SOLUTION

(1)
$$\frac{6xy^4 + 15x^3y}{3x^2y^2}$$
$$= \frac{6xy^4}{3x^2y^2} + \frac{15x^3y}{3x^2y^2}$$
$$= \frac{2y^2}{x} + \frac{5x}{y} \qquad \text{Answer}$$

PRACTICE PROBLEMS

Perform the following divisions.

(1) $\dfrac{6x^2+4x}{2x}$

(2) $\dfrac{xy^3-x^2y^2}{xy}$

(3) $\dfrac{8x^3y+6x^2y^2}{2xy}$

(4) $\dfrac{5x^2y^3-10xy^2}{5x^3y^2}$

Answers
(1) $3x+2$
(2) y^2-xy
(3) $4x^2+3xy$
(4) $\dfrac{y}{x}-\dfrac{2}{x^2}$

4.9 MULTIPLYING POLYNOMIALS BY POLYNOMIALS

The distributive law of multiplication, which we use to multiply polynomials by monomials, can also be extended to multiply polynomials by polynomials. In the examples below, the numbers indicate the order in which the terms are multiplied. Notice that when multiplying polynomials by polynomials (the second example), each term in the right polynomial is multiplied by each term in the left polynomial.

Monomial(Polynomial)

$$a(c+d+e)=ac+ad+ae$$

(Polynomial)(Polynomial)

$$(a+b)(c+d+e)=ac+ad+ae+bc+bd+be$$

SAMPLE PROBLEM

Problem 17: Multiply $(2x+3y)(x-5y)$.

PROCEDURE	SOLUTION
(1) Multiply each term in the right parentheses by $2x$ and $3y$. Combine like terms.	(1) $(2x+3y)(x-5y)$ $=2x(x)+2x(-5y)+3y(x)+3y(-5y)$ $=2x^2-10xy+3xy-15y^2$ $=2x^2-7xy-15y^2$ *Answer*

PRACTICE PROBLEMS

Answers
(1) $10ax-2ay+15bx-3by$
(2) $3x^2+7xy+2y^2$
(3) $x^3+5x^2+2x-12$
(4) x^6+2x^3-15

Perform the following multiplications.

(1) $(2a+3b)(5x-y)$
(2) $(x+2y)(3x+y)$
(3) $(x+3)(x^2+2x-4)$
(4) $(x^3-3)(x^3+5)$

5

5.1 SOLVING EQUATIONS

An **equation** is a mathematical statement that two quantities are equal. Equations which contain only numbers are either **true** or **false**. For example, the equation $3+4=7$ is true, while the equation $3+4=8$ is false. Equations which contain one or more variables are neither true nor false, but **open**. That is, their truth cannot be determined until all the variables are replaced by specific numerical values. For example, the equation $5x+4=14$ is an open equation which is true if x is replaced by 2, and which is false if x is replaced by 3.

$5x+4=14$ open
$5(2)+4=14$ true
$5(3)+4=18$ false

The specific values of the variables which make an open equation true are called the **roots** or **solutions** of the equation. The process of finding the roots is called **solving** the equation.

The most fundamental method of solving an equation is that of trial and error. After test values are chosen for the variables and substituted into the equation, each side of the equation is evaluated separately. If both sides result in the same value, the numbers chosen are roots. If not, new values are chosen and the process is repeated.

Solving an Equation by Trial and Error

Solve: $5x-2=3x+8$

Try $x=3$

$$5x-2=3x+8$$
$$5(3)-2 \stackrel{?}{=} 3(3)+8$$
$$15-2 \stackrel{?}{=} 9+8$$
$$13=17 \qquad \textit{False}$$

Try $x=4$

$$5x-2=3x+8$$
$$5(4)-2 \stackrel{?}{=} 3(4)+8$$
$$20-2 \stackrel{?}{=} 12+8$$
$$18=20 \qquad \textit{False}$$

Try $x=5$

$$5x-2=3x+8$$
$$5(5)-2 \stackrel{?}{=} 3(5)+8$$
$$25-2 \stackrel{?}{=} 15+8$$
$$23 \stackrel{?}{=} 23 \qquad \textit{True}$$

$x=5$ is a root.

This method of solving an equation is obviously not very efficient, and except in the most simple cases, will probably not work at all. Instead, other, more systematic methods have been developed which will now be discussed.

5.2 SOLVING LINEAR EQUATIONS IN ONE VARIABLE

In equations which contain only one variable, the highest power of the variable is called the equation's **degree**. For example, $5x-2=3x+8$ is called a **first degree**, or **linear equation**, $x^2+2x=10-x$ is called a **second degree**, or **quadratic equation**, and $x^3+5=x^2+6$ is called a **third degree**, or **cubic equation**.

The particular method used to solve an equation depends upon the equation's degree. For first degree (linear) equations, we use the method of **inverse operations**. The idea behind this method is to transform the given equation into an equivalent equation (an equation having the same roots) in which the variable appears alone on one side of the equation, and a number appears alone on the other. This number will be the root of the given equation.

In order to get the variable alone on one side of the equation, we must perform inverse operations to "undo" the operations on that side. Remember, addition undoes subtraction (and vice-versa), and multiplication undoes division (and vice-versa).

Since we want the new, transformed equation to have the same roots as the given equation, we must follow the equivalence principle stated below.

$$A = B$$
$$A+C = B+C$$
$$A-C = B-C$$
$$C \cdot A = C \cdot B$$
$$\frac{A}{C} = \frac{B}{C}$$

Equivalence Principle

Whenever a number is added to, subtracted from, multiplied by, or divided into one side of an equation, the same thing must be done on the other side of the equation.

In each of the examples below, the equation contains one operation, and thus is solved in one step (one inverse operation).

Undo Addition	Undo Subtraction	Undo Multiplication	Undo Division
$x+5= 9$	$x-3= 2$	$4x=12$	$\frac{x}{7}=2$
$-5 \quad -5$	$+3 \quad +3$	$\frac{\cancel{4}x}{\cancel{4}}=\frac{12}{4}$	$\cancel{7} \cdot \frac{x}{\cancel{7}}=7 \cdot 2$
$x \quad = 4$	$x \quad = 5$		
$\boxed{x=4}$	$\boxed{x=5}$	$\boxed{x=3}$	$\boxed{x=14}$

To solve equations containing more than one operation, we perform the inverse operations one step at a time. Although the order of performing the inverse operations is arbitrary, it is usually preferable to undo additions and subtractions before undoing multiplications and divisions. In each of the sample problems that follow, we will check the root we obtain by substituting it back into the original equation.

SAMPLE PROBLEMS

Problem 18: Solve $3x+8=23$ and check the root.

PROCEDURE	SOLUTION
(1) Subtract 8 from both sides of the equation. (Notice that if we divide both sides of the equation by 3 first, we get the fraction $\frac{23}{3}$ on the right side of the equation, which is more difficult to work with.)	(1) $\begin{array}{r} 3x+8=23 \\ -8 \quad -8 \\ \hline 3x \quad =15 \end{array}$
(2) Divide both sides by 3.	(2) $\dfrac{\cancel{3}x}{\cancel{3}} = \dfrac{15}{3}$ $x=5$ Answer
(3) Check the root by substituting it back into the original equation.	(3) Check $3x+8=23$ $3(5)+8 \stackrel{?}{=} 23$ $15+8 \stackrel{?}{=} 23$ $23 \stackrel{\checkmark}{=} 23$

Problem 19: Solve $10 = \dfrac{x}{3} - 1$ and check the root.

PROCEDURE	SOLUTION
(1) Add 1 to both sides of the equation.	(1) $\begin{array}{r} 10 = \dfrac{x}{3} - 1 \\ +1 \quad\quad +1 \\ \hline 11 = \dfrac{x}{3} \end{array}$
(2) Multiply both sides by 3.	(2) $3 \cdot 11 = \dfrac{x}{\cancel{3}} \cdot \cancel{3}$ $33 = x$ Answer
(3) Check the root by substitution.	(3) Check $10 = \dfrac{x}{3} - 1$ $10 \stackrel{?}{=} \dfrac{(33)}{3} - 1$ $10 \stackrel{?}{=} 11 - 1$ $10 \stackrel{\checkmark}{=} 10$

Problem 20: Solve $19 + 5x = 4$ and check the root.

PROCEDURE	SOLUTION
(1) Subtract 19 from both sides of the equation.	(1) $\begin{aligned} 19 + 5x &= 4 \\ -19 &= -19 \\ \hline 5x &= -15 \end{aligned}$
(2) Divide both sides by 5.	(2) $\dfrac{\cancel{5}x}{\cancel{5}} = \dfrac{-15}{5}$ $x = -3$ *Answer*
(3) Check the root by substitution.	(3) Check $19 + 5x = 4$ $19 + 5(-3) \stackrel{?}{=} 4$ $19 - 15 \stackrel{?}{=} 4$ $4 \stackrel{\checkmark}{=} 4$

Problem 21: Solve $-2 = 22 - 4x$ and check the root.

PROCEDURE	SOLUTION
(1) Subtract 22 from both sides of the equation.	(1) $\begin{aligned} -2 &= 22 - 4x \\ -22 &= -22 \\ \hline -24 &= -4x \end{aligned}$
(2) Divide both sides by -4. (Note that if we divide both sides by 4, instead, we get the equation $-6 = -x$. Then, by changing the signs on both sides, we get the same result, $6 = x$.)	(2) $\dfrac{-24}{-4} = \dfrac{\cancel{-4}x}{\cancel{-4}}$ $6 = x$ *Answer*
(3) Check the root by substitution.	(3) Check $-2 = 22 - 4x$ $-2 \stackrel{?}{=} 22 - 4(6)$ $-2 \stackrel{?}{=} 22 - 24$ $-2 \stackrel{\checkmark}{=} -2$

PRACTICE PROBLEMS

Solve each of the following equations for x.

(1) $x+8=3$
(2) $-5=x-9$
(3) $-7x=-21$
(4) $-4=\dfrac{x}{3}$
(5) $7x+11=32$
(6) $-6=\dfrac{x}{2}+1$
(7) $3=16+2x$
(8) $-7-3x=-4$

Answers
(1) -5
(2) 4
(3) 3
(4) -12
(5) 3
(6) -14
(7) $-6\frac{1}{2}$
(8) -1

5.3 SOLVING EQUATIONS CONTAINING PARENTHESES

To solve equation containing parentheses, first eliminate the parentheses by using the distributive law of multiplication.

Distributive Law
$A(B+C)=AB+AC$

SAMPLE PROBLEMS

Problem 22: Solve $3(2x-5)=27$.

PROCEDURE

(1) Using the distributive law, multiply each term inside the parentheses by 3.

(2) Add 15 to both sides of the equation.

(3) Divide both sides by 6.

SOLUTION

(1) $3(2x-5)=27$
$6x-15=27$

(2) $\begin{array}{r} 6x-15=27 \\ +15 +15 \\ \hline 6x = 42 \end{array}$

(3) $\dfrac{6x}{6}=\dfrac{42}{6}$
$x=7 \qquad$ *Answer*

Problem 23: Solve $5-2(3x-1)=25$.

PROCEDURE

(1) Multiply each term inside the parentheses by -2.

(2) Combine the numbers on the left side of the equation, and then subtract 7 from both sides.

(3) Divide both sides by -6.

SOLUTION

(1) $5-2(3x-1)=25$
$5-6x+2=25$

(2) $\begin{array}{r} 7-6x=25 \\ -7 -7 \\ \hline -6x=18 \end{array}$

(3) $\dfrac{-6x}{-6}=\dfrac{18}{-6}$
$x=-3 \qquad$ *Answer*

PRACTICE PROBLEMS

Solve each of the following equations for x.

(1) $6(3x+2)=30$
(2) $3+2(x-9)=6$
(3) $-10=5(4-2x)$
(4) $22=6+2(7-x)$

Answers			
(1)	1	(2)	$10\frac{1}{2}$
(3)	3	(4)	-1

5.4 SOLVING EQUATIONS CONTAINING LIKE VARIABLE TERMS

In some equations, the variable will appear in more than one term. If these terms are on the *same side* of the equation, we *combine* them. If they are on *opposite sides* of the equation, we use inverse operations to *eliminate* them from one side:

Same Side	Opposite Sides
$\underbrace{5x+3x}=24$	$5x = 3x+24$
$8x=24$	$-3x-3x$
$x=3$	$2x = 24$
	$x=12$

SAMPLE PROBLEMS

Problem 24: Solve $2(x-3)+3x=14$.

PROCEDURE

(1) Using the distributive law, eliminate the parentheses.

(2) Combine the like variable terms, and then add 6 to both sides of the equation.

(3) Divide both sides by 5.

SOLUTION

(1) $2(x-3)+3x=14$
$2x-6+3x=14$

(2) $5x-6=14$
$+6+6$
$5x=20$

(3) $\dfrac{\cancel{5}x}{\cancel{5}}=\dfrac{20}{5}$
$x=4$ *Answer*

Problem 25: Solve $5x-12=3x+4$.

PROCEDURE

(1) To eliminate the term $3x$ on the right side of the equation, subtract $3x$ from both sides of the equation. (Note that we could eliminate the term $5x$ instead, by subtracting $5x$ from both sides of the equation. It makes no difference.)

SOLUTION

(1) $5x-12=3x+4$
$-3x-3x$
$2x-12=4$

(2)	Add 12 to both sides of the equation.	(2)	$\begin{aligned} 2x - 12 &= 4 \\ +12 & +12 \\ \hline 2x &= 16 \end{aligned}$
(3)	Divide both sides by 2.	(3)	$\dfrac{\cancel{2}x}{\cancel{2}} = \dfrac{16}{2}$ $x = 8$ Answer

Problem 26: Solve $4x - 3 = 15 - 5x$.

	PROCEDURE		SOLUTION
(1)	Add $5x$ to both sides of the equation.	(1)	$\begin{aligned} 4x - 3 &= 15 - 5x \\ +5x & +5x \\ \hline 9x - 3 &= 15 \end{aligned}$
(2)	Add 3 to both sides.	(2)	$\begin{aligned} 9x - 3 &= 15 \\ +3 & +3 \\ \hline 9x &= 18 \end{aligned}$
(3)	Divide both sides by 9.	(3)	$\dfrac{\cancel{9}x}{\cancel{9}} = \dfrac{18}{9}$ $x = 2$ Answer

PRACTICE PROBLEMS

Solve each of the following equations for x.

(1) $5(x - 4) - 2x = 3$ (2) $2x - 6 = 7x + 14$
(3) $15 - 3x = x + 3$ (4) $17 - 2x = 9 - 6x$

> Answers
> (1) $7\frac{2}{3}$ (2) -4
> (3) 3 (4) -2

5.5 SOLVING EQUATIONS CONTAINING FRACTIONS

   To solve equations containing fractions, it is usually preferable to eliminate the fractions first. This is done by multiplying both sides of the equation by a common denominator of the fractions in the equation. All the denominators will cancel, leaving an equation free of fractions. This procedure is sometimes referred to as "clearing" the equation of fractions.

SAMPLE PROBLEMS

Problem 27: Solve $\dfrac{x}{2} + \dfrac{x}{3} = 10$.

PROCEDURE	SOLUTION
(1) Multiply both sides of the equation by 6, a common denominator of $\frac{x}{2}$ and $\frac{x}{3}$. Cancel the denominators.	(1) $\dfrac{x}{2} + \dfrac{x}{3} = 10$ $6\left(\dfrac{x}{2} + \dfrac{x}{3}\right) = 6(10)$ $\overset{3}{\cancel{6}} \cdot \dfrac{x}{\cancel{2}} + \overset{2}{\cancel{6}} \cdot \dfrac{x}{\cancel{3}} = 60$ $3x + 2x = 60$
(2) Combine the like terms, and divide both sides of the equation by 5.	(2) $5x = 60$ $\dfrac{\cancel{5}x}{\cancel{5}} = \dfrac{60}{5}$ $x = 12 \qquad$ Answer

Problem 28: Solve $\dfrac{4x}{5} = \dfrac{x}{2} + 6$.

PROCEDURE	SOLUTION
(1) Multiply both sides of the equation by 10, a common denominator of $\frac{4x}{5}$ and $\frac{x}{2}$. Cancel the denominators.	(1) $\dfrac{4x}{5} = \dfrac{x}{2} + 6$ $10\left(\dfrac{4x}{5}\right) = 10\left(\dfrac{x}{2} + 6\right)$ $\overset{2}{\cancel{10}} \cdot \dfrac{4x}{\cancel{5}} = \overset{5}{\cancel{10}} \cdot \dfrac{x}{\cancel{2}} + 10 \cdot 6$ $8x = 5x + 60$
(2) Subtract $5x$ from both sides of the equation.	(2) $\begin{array}{r} 8x = 5x + 60 \\ -5x -5x \\ \hline 3x = 60 \end{array}$
(3) Divide both sides by 3.	(3) $\dfrac{\cancel{3}x}{\cancel{3}} = \dfrac{60}{3}$ $x = 20 \qquad$ Answer

Problem 29: Solve $\dfrac{3}{x} + \dfrac{3}{4x} = \dfrac{5}{8}$.

PROCEDURE	SOLUTION
(1) Multiply both sides of the equation by $8x$, a common denominator of all the fractions. Cancel the denominators.	(1) $\dfrac{3}{x} + \dfrac{3}{4x} = \dfrac{5}{8}$ $8x\left(\dfrac{3}{x} + \dfrac{3}{4x}\right) = 8x\left(\dfrac{5}{8}\right)$ $8\cancel{x}\cdot\dfrac{3}{\cancel{x}} + \cancel{8}x\cdot\dfrac{3}{\cancel{4x}}^{2} = \cancel{8}x\cdot\dfrac{5}{\cancel{8}}$ $24 + 6 = 5x$ $30 = 5x$
(2) Divide both sides by 5.	(2) $\dfrac{30}{5} = \dfrac{\cancel{5}x}{\cancel{5}}$ $6 = x$ Answer

PRACTICE PROBLEMS

Solve each of the following equations for x.

(1) $\dfrac{2x}{9} + \dfrac{1}{3} = 5$ (2) $\dfrac{x}{5} - \dfrac{x}{3} = 1$

(3) $\dfrac{2x}{3} - 2 = \dfrac{3x}{4}$ (4) $\dfrac{2}{x} = \dfrac{9}{4} - \dfrac{5}{2x}$

Answers	
(1) 21	(2) $-7\frac{1}{2}$
(3) -24	(4) 2

5.6 SOLVING PROPORTIONS

As you recall, a proportion is a statement that two fractions are equal. To solve a proportion, cross multiply the numerators and denominators of the fractions, set the products equal, and then solve the resulting equation.

$\dfrac{2x}{6} = \dfrac{4}{3}$

$3 \cdot 2x = 6 \cdot 4$

$6x = 24$

$x = 4$

SAMPLE PROBLEM

Problem 30: Solve $\dfrac{x+6}{2} = \dfrac{4x}{5}$.

	PROCEDURE		SOLUTION
(1)	Cross multiply the numerators and denominators and set them equal.	(1)	$\dfrac{x+6}{2} = \dfrac{4x}{5}$ $5(x+6) = 2(4x)$ $5x + 30 = 8x$
(2)	Subtract $5x$ from both sides of the equation.	(2)	$\begin{array}{r}5x+30 = 8x\\ -5x -5x\\ \hline 30 = 3x\end{array}$
(3)	Divide both sides by 3.	(3)	$\dfrac{30}{3} = \dfrac{\cancel{3}x}{\cancel{3}}$ $10 = x$ Answer

PRACTICE PROBLEMS

Solve each of the following proportions for x.

(1) $\dfrac{2x}{9} = \dfrac{4}{3}$ (2) $\dfrac{x-1}{3} = \dfrac{2x}{5}$

(3) $\dfrac{5x}{6} = \dfrac{4x-3}{3}$ (4) $\dfrac{2x+3}{5} = \dfrac{3x-1}{4}$

Answers
(1) 6   (2) −5
(2) 2   (4) $2\tfrac{3}{7}$

5.7 SOLVING EQUATIONS CONTAINING SQUARE ROOTS

$\begin{aligned}\sqrt{x} + 3 &= 9\\ -3 & -3\\ \hline \sqrt{x} &= 6\\ (\sqrt{x})^2 &= (6)^2\\ x &= 36\end{aligned}$

To solve equations in which the variable appears underneath a square root sign, use inverse operations to get the square root alone on one side of the equation, and then square both sides. Since squaring is the inverse of taking a square root $\left((\sqrt{x})^2 = x\right)$, the square root sign will be eliminated from the equation.

SAMPLE PROBLEM

Problem 31: Solve $\sqrt{3x+1} + 2 = 6$ and check the root.

PROCEDURE	SOLUTION
(1) Subtract 2 from both sides.	(1) $\begin{aligned} \sqrt{3x+1} + 2 &= 6 \\ -2 & -2 \\ \hline \sqrt{3x+1} &= 4 \end{aligned}$
(2) Square both sides.	(2) $(\sqrt{3x+1})^2 = (4)^2$ $3x+1 = 16$
(3) Solve the resulting equation for x.	(3) $\begin{aligned} 3x+1 &= 16 \\ -1 & -1 \\ \hline 3x &= 15 \\ \frac{\cancel{3}x}{\cancel{3}} &= \frac{15}{3} \\ x &= 5 \quad \text{Answer} \end{aligned}$
(4) Check the root by substituting it into the original equation.	(4) Check $\sqrt{3x+1} + 2 = 6$ $\sqrt{3(5)+1} + 2 \stackrel{?}{=} 6$ $\sqrt{15+1} + 2 \stackrel{?}{=} 6$ $\sqrt{16} + 2 \stackrel{?}{=} 6$ $4 + 2 \stackrel{?}{=} 6$ $6 \stackrel{\checkmark}{=} 6$

Sometimes the root we obtain by this procedure will solve the "squared equation" (the equation after squaring both sides), but will *not* solve the original equation.

Solve: $\sqrt{2x} + 5 = 1$

Solution	Check
$\begin{aligned} \sqrt{2x} + 5 &= 1 \\ -5 & -5 \\ \hline \sqrt{2x} &= -4 \end{aligned}$ $(\sqrt{2x})^2 = (-4)^2$ $2x = 16$ $x = 8$	$\sqrt{2x} + 5 = 1$ $\sqrt{2 \cdot 8} + 5 \stackrel{?}{=} 1$ $\sqrt{16} + 5 \stackrel{?}{=} 1$ $4 + 5 \stackrel{?}{=} 1$ $9 \neq 1$

As you can see, the root $x=8$ solves the squared equation, $2x=16$, but does not solve the original equation, $\sqrt{2x}+5=1$. In cases like this, we call the root an **extraneous root** (false root) and discard it.

PRACTICE PROBLEMS

Solve each of the following equations for x, and check the root found.

(1) $\sqrt{3x-12}+2=5$

(2) $4\sqrt{\dfrac{2x}{3}}=24$

(3) $\sqrt{7x}=\sqrt{2x+7}$

(4) $\sqrt{2x}+8=2$

Answers
(1) 7 (2) 54
(3) $1\frac{2}{5}$ (4) no roots

5.8 SOLVING LITERAL EQUATIONS

Equations which contain more than one variable are called **literal equations** (equations with **letters**). All formulas, for example, are literal equations. To solve a literal equation for one of its variables means to express that variable in terms of all the others. This is accomplished by treating all the other variables as though they were simply numbers, and then solving for the desired variable in the usual manner. Consider the example below. Since we are solving for the variable x, we treat the other variables, a, b, and c as though they were numbers.

Solve for x: $ax+b=c$

$$
\begin{aligned}
ax+b &= c \\
-b & -b \quad \text{(subtract } b\text{)} \\
ax &= c-b \\
\dfrac{\cancel{a}x}{\cancel{a}} &= \dfrac{c-b}{a} \quad \text{(divide by } a\text{)} \\
x &= \dfrac{c-b}{a}
\end{aligned}
$$

The method of solving literal equations for one of its variables is particularly useful for changing the subject of a formula from one of its variables to another. In the example below, we use this method to change the subject of the formula $A=\dfrac{bh}{2}$ from A, the area of the triangle to h, the height of the triangle.

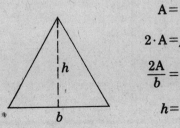

$$
\begin{aligned}
A &= \dfrac{bh}{2} \\
2 \cdot A &= \cancel{2} \cdot \dfrac{bh}{\cancel{2}} \quad \text{(multiply by 2)} \\
\dfrac{2A}{b} &= \dfrac{\cancel{b}h}{\cancel{b}} \quad \text{(divide by } b\text{)} \\
h &= \dfrac{2A}{b}
\end{aligned}
$$

SAMPLE PROBLEMS

Problem 32: Solve $5ax + 4a^2 = 2ax + 10a^2$ for x.

PROCEDURE	SOLUTION
(1) Subtract $2ax$ from both sides of the equation.	(1) $\begin{array}{r} 5ax+4a^2 = 2ax+10a^2 \\ -2ax \qquad -2ax \\ \hline 3ax+4a^2 = \qquad 10a^2 \end{array}$
(2) Subtract $4a^2$ from both sides of the equation.	(2) $\begin{array}{r} 3ax+4a^2 = 10a^2 \\ -4a^2 \quad -4a^2 \\ \hline 3ax = 6a^2 \end{array}$
(3) Divide both sides by $3a$.	(3) $\dfrac{3ax}{3a} = \dfrac{6a^2}{3a}$ $x = 2a$ *Answer*

Problem 33: In the formula for the perimeter of a rectangle, $P = 2(L+W)$, change the subject to W, the width.

$$P = 2(L+W)$$

PROCEDURE	SOLUTION
(1) Using the distributive law, eliminate the parentheses.	(1) $P = 2(L+W)$ $P = 2L + 2W$
(2) Subtract $2L$ from both sides of the equation.	(2) $\begin{array}{r} P \quad = 2L+2W \\ -2L \quad -2L \\ \hline P-2L = \quad 2W \end{array}$
(3) Divide both sides by 2.	(3) $\dfrac{P-2L}{2} = \dfrac{2W}{2}$ $W = \dfrac{P-2L}{2}$ *Answer*

PRACTICE PROBLEMS

Solve each of the following equations for x.

(1) $y = mx + b$

(2) $ax + by = c$

(3) $3ax + b^2 = 4b^2 - ax$

(4) $\dfrac{a}{x} + \dfrac{b}{3x} = 1$

Answers

(1) $\dfrac{y-b}{m}$

(2) $\dfrac{c-by}{a}$

(3) $\dfrac{3b^2}{4a}$

(4) $\dfrac{3a+b}{3}$

5.9 SOLVING EQUATIONS SIMULTANEOUSLY—ADDITION METHOD

To solve a linear equation containing two variables means to find a *pair* of numbers (one for each variable) which makes the equation true. Each pair is then called a root of the equation. For example, one root of the equation $x + y = 5$ is the pair $x = 1$, $y = 4$, and another root is the pair $x = 2$, $y = 3$.

Unlike linear equations in one variable, which have only one root, linear equations in two variables have an infinite number of roots.

Some Roots of $x+y=3$	Some Roots of $x-y=1$
$x=0, y=3$	$x=0, y=-1$
$x=1, y=2$	$x=1, y=0$
$x=2, y=1$	$x=2, y=1$
$x=3, y=0$	$x=3, y=2$
$x=4, y=-1$	$x=4, y=3$
etc.	etc.

As indicated by the dotted lines, the pair $x=2$, $y=1$ is a common root of both equations. This is an example of the fact that, except for a few special cases, two linear equations in two variables always have one common root. The exceptions will be discussed later.

The process of finding the common root of two equations in two variables is referred to as solving the equations **simultaneously**. There are two general methods for doing this. The idea behind the first method, called the **addition method**, is to eliminate one of the variables by adding the equations together. In our sample problem note that the equations used are the same as those in the preceding example.

SAMPLE PROBLEM

Problem 34: Find the common root of the equations $x+y=3$ and $x-y=1$. Check the root in both equations.

	PROCEDURE		SOLUTION
(1)	Write the equations with the like variables lined up and add them, thus eliminating the variable y.	(1)	$x+y=3$ $x-y=1$ $\overline{2x =4}$
(2)	Solve the resulting equation for x.	(2)	$\dfrac{\cancel{2}x}{\cancel{2}} = \dfrac{4}{2}$ $x=2$
(3)	Substitute this value for x into either one of the original equations.	(3)	$x+y=3$ or $x-y=1$ $2+y=3$ $2-y=1$
(4)	Solve the resulting equation for y.	(4)	$\begin{array}{r}2+y=3\\-2-2\\\hline y=1\end{array}$ or $\begin{array}{r}2-y=1\\-2-2\\\hline -y=-1\\y=1\end{array}$ $x=2, y=1$ Answer
(5)	Check the root by substituting it into both of the original equations.	(5)	Check $x+y=3$ $x-y=1$ $2+1 \overset{?}{=} 3$ $2-1 \overset{?}{=} 1$ $3 \overset{\checkmark}{=} 3$ $1 \overset{\checkmark}{=} 1$

If adding the two equations does not eliminate one of the variables directly (as in the preceding sample problem), then we must multiply either one or both of the equations by appropriate numbers so that one of the variables will have numerical coefficients which are opposite signed numbers (numbers with the same absolute values, but opposite signs).

SAMPLE PROBLEMS

Problem 35: Solve the equations $3x+y=7$ and $4x+3y=6$ simultaneously, and check the root in both equations.

	PROCEDURE		SOLUTION
(1)	Write the equations with the like variables lined up.	(1)	$3x+y=7$ $4x+3y=6$

Solving Linear Equations and Inequalities / 43

(2)	To eliminate the variable y, multiply the top equation by -3. Leave the bottom equation as it is.	(2)	$-3(3x+y=7) \rightarrow -9x-3y=-21$ $4x+3y=6 \rightarrow 4x+3y=6$
(3)	Add the two equations.	(3)	$-9x-3y=-21$ $\underline{4x+3y=6}$ $-5x=-15$
(4)	Solve the resulting equation for x.	(4)	$\dfrac{-\cancel{5}x}{-\cancel{5}} = \dfrac{-15}{-5}$ $x = 3$
(5)	Substitute this value for x into either one of the original equations, and solve the resulting equation for y.	(5)	$3x+y=7$ $\quad\left(\begin{array}{c}\text{top}\\ \text{equation}\end{array}\right)$ $3(3)+y=7$ $9+y=7$ $\underline{-9-9}$ $y=-2$ $x=3,\ y=-2$ $\qquad$ **Answer**
(6)	Check the root by substituting it into both of the original equations.	(6)	Check $\begin{array}{c|c} 3x+y=7 & 4x+3y=6 \\ 3(3)+(-2)\stackrel{?}{=}7 & 4(3)+3(-2)\stackrel{?}{=}6 \\ 9-2\stackrel{?}{=}7 & 12-6\stackrel{?}{=}6 \\ 7\stackrel{\checkmark}{=}7 & 6\stackrel{\checkmark}{=}6 \end{array}$

Problem 36: Solve the equations $2x = 3 - 3y$ and $2y = 13 - 5x$ simultaneously, and check the root in both equations.

	PROCEDURE		SOLUTION
(1)	Using inverse operations, rewrite the equations with the variables on one side and the numbers on the other. (This is called **standard form**.)	(1)	$2x = 3 - 3y$ $\qquad$ $2y = 13 - 5x$ $\underline{+3y+3y}$ $\qquad\qquad$ $\underline{+5x+5x}$ $2x+3y = 3$ $\qquad\qquad$ $5x+2y=13$
(2)	Write the resulting equations with the like variables lined up.	(2)	$2x+3y=3$ $5x+2y=13$
(3)	To eliminate the variable x (the choice is arbitrary), multiply the top equation by 5 (the coefficient of x in the bottom equation), and the bottom equation by -2 (the opposite of the coefficient of x in the top equation).	(3)	$5(2x+3y=3) \rightarrow 10x+15y=15$ $-2(5x+2y=13) \rightarrow -10x-4y=-26$

44 / *Algebra Simplified and Self-Taught*

(4) Add the two equations, and solve the resulting equation for y.

(4)
$$10x+15y= 15$$
$$-10x-4y=-26$$
$$\overline{11y=-11}$$
$$\frac{\cancel{11}y}{\cancel{11}}=\frac{-11}{11}$$
$$y=-1$$

(5) Substitute this value for y into either one of the original equations, and solve the resulting equation for x.

(5)
$$2x=3-3y$$
$$2x=3-3(-1) \quad \left(\begin{array}{c}\text{top}\\\text{equation}\end{array}\right)$$
$$2x=3+3$$
$$2x=6$$
$$\frac{\cancel{2}x}{\cancel{2}}=\frac{6}{2}$$
$$x=3$$
$$x=3, y=-1 \qquad \text{Answer}$$

(6) Check the root by substituting it into both of the original equations.

(6) Check
$$2x=3-3y \qquad\qquad 2y=13-5x$$
$$2(3)\stackrel{?}{=}3-3(-1) \qquad 2(-1)\stackrel{?}{=}13-5(3)$$
$$6\stackrel{?}{=}3+3 \qquad\qquad -2\stackrel{?}{=}13-15$$
$$6\stackrel{\checkmark}{=}6 \qquad\qquad -2\stackrel{\checkmark}{=}-2$$

PRACTICE PROBLEMS

Solve each of the following pairs of equations simultaneously.

(1) $2x+y=8$
     $x-y=1$

(2) $3x-2y=17$
     $2x+y=9$

(3) $5x+3y=28$
     $7x-2y=2$

(4) $3x=2y-18$
     $4y=8-x$

Answers
(1) $x=3, y=2$
(2) $x=5, y=-1$
(3) $x=2, y=6$
(4) $x=-4, y=3$

5.10 SOLVING EQUATIONS SIMULTANEOUSLY—SUBSTITUTION METHOD

A second method of solving equations simultaneously is called the **substitution method**. The idea behind this method is to eliminate one of the variables by substituting an expression given for it by one of the equations into the other equation. This method is particularly useful when one of the equations has a form, such as $y=2x+1$ or $x=3y-5$, in which one of the variables is expressed directly in terms of the other variable.

SAMPLE PROBLEM

Problem 37: Solve $y=2x-1$ and $x+3y=11$ simultaneously, and check the roots in both equations.

PROCEDURE	SOLUTION
(1) Substitute the expression given for y by the first equation ($y=2x-1$) into the second equation, thus eliminating the variable y.	(1) $y=2x-1$ $x+3y=11$ $x+3(2x-1)=11$
(2) Solve the resulting equation for x.	(2) $x+3(2x-1)=11$ $x+6x-3=11$ $7x-3=11$ $+3+3$ $7x=14$ $\dfrac{7x}{7}=\dfrac{14}{7}$ $x=2$
(3) To obtain y, substitute this value for x into the equation originally eliminated.	(3) $y=2x-1$ $y=2(2)-1$ $y=4-1$ $y=3$ $x=2,\ y=3$ Answer
(4) Check the root by substituting it into both of the original equations.	(4) Check $y=2x-1 \quad\bigg|\quad x+3y=11$ $(3)\overset{?}{=}2(2)-1 \quad\bigg|\quad (2)+3(3)\overset{?}{=}11$ $3\overset{?}{=}4-1 \quad\bigg|\quad 2+9\overset{?}{=}11$ $3\overset{\checkmark}{=}3 \quad\bigg|\quad 11\overset{\checkmark}{=}11$

PRACTICE PROBLEMS

Solve each of the following pairs of equations simultaneously.

(1) $y=3x$
    $x+2y=42$

(2) $x=2y$
    $4x-9y=3$

(3) $y=3x+1$
    $x+y=9$

(4) $x=2y-3$
    $5x-y=3$

Answers
(1) $x=6,\ y=18$
(2) $x=-6,\ y=-3$
(3) $x=2,\ y=7$
(4) $x=1,\ y=2$

As previously noted, there are a few special cases in which two linear equations in two variables do *not* have one common root. These special cases fall into two categories—inconsistent equations and dependent equations.

Inconsistent equations are equations, such as $x+y=3$ and $x+y=4$, which are impossible to solve simultaneously (the sum of x and y cannot be both 3 and 4 at the same time), and thus have *no roots* in common.

Dependent equations are equations, such as $x+y=5$ and $2x+2y=10$, which are not really different from each other (one equation is simply a multiple of the other), and thus have *all their roots* in common.

5.11 SOLVING LINEAR INEQUALITIES IN ONE VARIABLE

An **inequality** is a mathematical statement in which two quantities are related by one of the following signs: $>$ (greater than), $\geq$ (greater than or equal to), $<$ (less than), or $\leq$ (less than or equal to).

Linear inequalities in one variable, like linear equations in one variable, are solved by the method of inverse operations. However, when solving inequalities, one additional principle must be observed.

Inequality Principle

When *multiplying* or *dividing* both sides of an inequality by a *negative number*, the direction of the inequality sign must be reversed.

This principle is demonstrated below. Notice that when we multiply or divide by 2, the direction of the inequality sign remains the same, but that when we multiply or divide by -2, the direction of the inequality sign must be reversed in order to make the resulting inequality true.

Multiply by 2	*Divide by 2*	*Multiply by -2*	*Divide by -2*
$8>6$	$8>6$	$8>6$	$8>6$
$2(8)>2(6)$	$\dfrac{8}{2}>\dfrac{6}{2}$	$-2(8)<-2(6)$	$\dfrac{8}{-2}<\dfrac{6}{-2}$
$16>12$	$4>3$	$-16<-12$	$-4<-3$

After an inequality is solved, its roots can be displayed graphically on a number line in the following way:

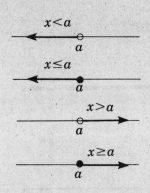

> **To Graph the Roots of an Inequality:**
>
> (1) At the point representing the upper or lower limit of the roots, draw a small circle. If this number is included in the roots, blacken in the circle; if not, leave the circle empty.
>
> (2) If all the other roots are less than this number, draw an arrow from the circle to the left. If all the other roots are greater than this number, draw an arrow from the circle to the right.

SAMPLE PROBLEMS

Problem 38: Solve $2x-5<3$, and represent the roots on a number line.

PROCEDURE	SOLUTION
(1) Add 5 to both sides of the inequality.	(1) $\begin{array}{r} 2x-5<3 \\ +5+5 \\ \hline 2x<8 \end{array}$
(2) Divide both sides by 2.	(2) $\dfrac{\cancel{2}x}{\cancel{2}} < \dfrac{8}{2}$ $x<4$ **Answer**
(3) On a number line, draw an empty circle at 4. Draw an arrow from the circle to the left.	(3) number line with open circle at 4, arrow left, showing $-1,0,1,2,3,4$

Problem 39: Solve $\dfrac{x}{6} - \dfrac{x}{2} \leq 1$, and represent the roots on a number line.

PROCEDURE	SOLUTION
(1) Multiply both sides of the inequality by 6, a common denominator of $\frac{x}{6}$ and $\frac{x}{2}$. Cancel the denominators, and combine like terms.	(1) $\dfrac{x}{6} - \dfrac{x}{2} \leq 1$ $6\left(\dfrac{x}{6} - \dfrac{x}{2}\right) \leq 6(1)$ $\cancel{6} \cdot \dfrac{x}{\cancel{6}} - \cancel{6}^{3} \cdot \dfrac{x}{\cancel{2}} \leq 6$ $x - 3x \leq 6$ $-2x \leq 6$ $\downarrow$
(2) Divide both sides by -2 and reverse the direction of the inequality sign (dividing by a negative number).	(2) $\dfrac{-\cancel{2}x}{-\cancel{2}} \geq \dfrac{6}{-2}$ $x \geq -3$ **Answer**

(3) On a number line, draw a blackened circle at -3. Draw an arrow from the circle to the right.

(3) [number line with blackened circle at -3 and arrow extending to the right, showing points $-3, -2, -1, 0, 1$]

PRACTICE PROBLEMS

Solve each of the following inequalities for x.

(1) $3x - 4 > 14$

(2) $12 - 2x \geq 4$

(3) $6x + 1 < 3(x - 2)$

(4) $\dfrac{x}{3} - \dfrac{x}{2} \geq 5$

Answers
(1) $x > 6$
(2) $x \leq 4$
(3) $x < -2\frac{1}{3}$
(4) $x \leq -30$

6

6.1 FACTORING

Multiplication
$\longrightarrow$
$5 \cdot 4 = 20$
$\longleftarrow$
Factoring

As you recall, the numbers used in multiplication are called **factors**, and the result is called the **product**. For example, in the statement $5 \cdot 4 = 20$, 5 and 4 are factors, and 20 is the product. If we reverse this statement and write $20 = 5 \cdot 4$, we say that we have "factored" 20 into 5 times 4. In other words, **factoring** is the process of writing a number, or algebraic expression, as a product of factors.

The particular method used to factor an algebraic expression is based on the type of expression that is given. We will examine methods for factoring three types of expressions: (1) polynomials having a common monomial factor, (2) trinomials of the general form $ax^2 + bx + c$, and (3) binomials of the general form $A^2 - B^2$, called the difference of two squares.

6.2 FACTORING POLYNOMIALS HAVING A COMMON MONOMIAL FACTOR

Distributive Law
$A(B+C) = AB + AC$

To multiply a polynomial by a monomial, we use the distributive law and multiply each term in the polynomial by the monomial:

$$5a(x + y^3 - z^2) = 5ax + 5ay^3 - 5az^2$$

The monomial, $5a$, is a common factor in each of the terms of the resulting polynomial. Therefore, to factor a polynomial whose terms have a common monomial factor, we use the distributive law in reverse, and write the polynomial as the product of the common monomial factor and another polynomial. That is,

$$5ax + 5ay^3 - 5az^2 = 5a(x + y^3 - z^2)$$

This process is sometimes referred to as "factoring out" the common monomial factor.

In general, to factor out a common monomial factor from a polynomial, use the following procedure.

> **To Factor a Polynomial Having a Common Monomial Factor:**
> (1) Write the common monomial factor next to a pair of parentheses.
> (2) Divide this factor into each of the terms of the given polynomial, and write the resulting terms inside the parentheses.

$$6x^3 + 10xy^2$$
$$= 2x(\qquad)$$
$$= 2x(3x^2 + 5y^2)$$
$$\boxed{\tfrac{6x^3}{2x}} + \boxed{\tfrac{10xy^2}{2x}}$$

After completing this procedure, two checks should be made. First, multiply the common monomial factor by the polynomial inside the parentheses to verify that their product is the given polynomial. Second, look at the polynomial inside the parentheses to see if any common factors remain. If so, repeat the procedure and factor them out. If not, then the **greatest common factor** (g.c.f) has been factored out. For example, in the expression shown opposite, after factoring out the common monomial factor $3x$, the polynomial inside the parentheses, $x + 2x^2y$, still has a common factor of x. After factoring out x, no common factors remain, and thus $3x^2$, the greatest common factor, has been factored out.

$$3x^2 + 6x^3y$$
$$= 3x(x + 2x^2y)$$
$$= 3x^2(1 + 2xy)$$
$$\uparrow$$
$$\text{g.c.f.}$$

SAMPLE PROBLEM

Problem 40: Factor $6x^2y^2 + 9xy^3 - 3xy^2$.

PROCEDURE	SOLUTION
(1) By looking at the numerical and variable factors of each of the terms, notice that $3xy^2$ is a common factor. Write this factor next to a pair of parentheses.	(1) $6x^2y^2 + 9xy^3 - 3xy^2$ $= 3xy^2(\qquad)$
(2) Divide $3xy^2$ into each of the terms of the given polynomial, and write the resulting terms inside the parentheses.	(2) $= 3xy^2(2x + 3y - 1)$ Answer $\boxed{\tfrac{6x^2y^2}{3xy^2}} + \boxed{\tfrac{9xy^3}{3xy^2}} - \boxed{\tfrac{3xy^2}{3xy^2}}$

PRACTICE PROBLEMS

Factor out the common monomial factor from each of the following polynomials.

(1) $2a^2x + 4ab$
(2) $6x^2y^3 - 9x^2y$
(3) $5xy^2 + 10x^2y^2 - 15xy^3$
(4) $a^2b^2 - 3a^5b^3 - a^2b$

> **Answers**
> (1) $2a(ax + 2b)$
> (2) $3x^2y(2y^2 - 3)$
> (3) $5xy^2(1 + 2x - 3y)$
> (4) $a^2b(b - 3a^3b^2 - 1)$

6.3 FACTORING TRINOMIALS OF THE GENERAL FORM $ax^2 + bx + c$

One of the methods used to multiply binomials is called the FOIL method (First, Outer, Inner, Last). As illustrated in the example, the first term in the result is obtained by multiplying the First terms in the binomials, A and C, the second term in the result is obtained by multiplying the Outer terms in the binomials, A and D, the third term in the result is obtained by multiplying the Inner terms of the binomials, B and C, and the fourth term in the result is obtained by multiplying the Last terms in the binomials, B and D.

The FOIL Method of Multiplying Binomials

$$(A+B)(C+D) = \underset{\text{First}}{AC} + \underset{\text{Outer}}{AD} + \underset{\text{Inner}}{BC} + \underset{\text{Last}}{BD}$$

When we use the FOIL method to multiply binomials, such as $(3x+1)$ and $(2x+5)$, whose First terms are like variable terms, and whose Last terms are numbers, we find that the Outer and Inner products always combine, leaving a trinomial of the general form $ax^2 + bx + c$, where a, b, and c are specific signed numbers. Consider the following examples.

$$(3x+1)(2x+5) = \underset{F}{6x^2} + \underset{O}{15x} + \underset{I}{2x} + \underset{L}{5}$$
$$= 6x^2 + 17x + 5$$

$$(x-6)(x+4) = \underset{F}{x^2} + \underset{O}{4x} - \underset{I}{6x} - \underset{L}{24}$$
$$= x^2 - 2x - 24$$

$$(2x-3)(x-2) = \underset{F}{2x^2} - \underset{O}{4x} - \underset{I}{3x} + \underset{L}{6}$$
$$= 2x^2 - 7x + 6$$

Therefore, to factor a trinomial of the form ax^2+bx+c, we must reverse the FOIL procedure and write the trinomial as the product of two binomial factors of the form $(mx+n)$ and $(rx+s)$, where m, n, r, and s are also specific signed numbers. That is,

$$ax^2+bx+c=(mx+n)(rx+s)$$

In general, to find the two binomial factors, $(mx+n)$ and $(rx+s)$, we use this procedure:

To Factor Trinomials of the Form ax^2+bx+c Having Binomial Factors of the Form $(mx+n)$ and $(rx+s)$, where a, b, c, m, n, r, and s, are specific signed numbers:

(1) As the First terms of two binomial factors, write all possible terms, mx and rx, whose product is ax^2, the first term of the given trinomial.
(2) As the Last terms of the binomial factors, write all possible pairs of numbers, n and s, whose product is c, the third (last) term of the given trinomial.
(3) Combine the Outer and Inner products of all the possible binomial factors, and choose the combination that gives bx, the second (middle) term of the given trinomial.

SAMPLE PROBLEMS

Problem 41: Factor x^2-5x+6.

PROCEDURE	SOLUTION
(1) Write x as the First term of two binomial factors.	(1) $(\overset{\frown{x^2}}{x\quad})(x\quad)$
(2) As the Last terms of the two binomials, write all possible pairs of numbers whose product is $+6$, the last term of the given trinomial. Note that the signs of the numbers must be the same, both $+$ or both $-$.	(2) $(x+1)\overset{\frown{+6}}{(x+6)}$ $(x-1)\overset{\frown{+6}}{(x-6)}$ $(x+2)\overset{\frown{+6}}{(x+3)}$ $(x-2)\overset{\frown{+6}}{(x-3)}$

Factoring Polynomials / 53

(3) Combine the Outer and Inner products of each possibility, and choose the combination that gives $-5x$, the middle term of the given trinomial.

(3)
$(x+1)(x+6)$ with outer $+6x$, inner $+x$: $+6x+x=+7x$

$(x-1)(x-6)$ with outer $-6x$, inner $-x$: $-6x-x=-7x$

$(x+2)(x+3)$ with outer $+3x$, inner $+2x$: $+3x+2x=+5x$

$(x-2)(x-3)$ with outer $-3x$, inner $-2x$: $-3x-2x=\boxed{-5x}$

$x^2-5x+6=(x-2)(x-3)$ Answer

Problem 42: Factor x^2+4x-5.

PROCEDURE

(1) Write x as the First term of two binomial factors.

(2) As the Last terms of the two binomials, write all possible pairs of numbers whose product is -5, the last term of the given trinomial. Note that the signs of the numbers must be different, one + and one −.

(3) Combine the Outer and Inner products of each possibility, and choose the combination that gives $+4x$, the middle term of the given trinomial.

SOLUTION

(1) $(x \quad)(x \quad)$ with x^2

(2)
$(x+1)(x-5)$ with -5

$(x-1)(x+5)$ with -5

(3)
$(x+1)(x-5)$ outer $-5x$, inner $+x$: $-5x+x=-4x$

$(x-1)(x+5)$ outer $+5x$, inner $-x$: $+5x-x=\boxed{+4x}$

$x^2+4x-5=(x-1)(x+5)$ Answer

Problem 43: Factor $x^2-8x+12$.

PROCEDURE

(1) Write x as the First term of two binomial factors.

(2) As the Last terms of the two binomials, write all possible pairs of numbers whose product is $+12$, the last term of the given trinomial. Note that the signs of the numbers must be the same, both + or both −.

SOLUTION

(1) $(x \quad)(x \quad)$ with x^2

(2)
$(x+1)(x+12)$
$(x-1)(x-12)$
$(x+2)(x+6\ )$
$(x-2)(x-6\ )$
$(x+3)(x+4\ )$
$(x-3)(x-4\ )$

54 / Algebra Simplified and Self-Taught

(3) Combine the Outer and Inner products of each possibility, and choose the combination that gives $-8x$, the middle term of the given trinomial. (Note that only the correct combination is shown.)

(3) $(x-2)(x-6)$ with outer $-6x$, inner $-2x$: $-6x-2x = \boxed{-8x}$

$$x^2 - 8x + 12 = (x-2)(x-6) \quad \text{Answer}$$

Problem 44: Factor $2x^2 + 5x + 3$.

PROCEDURE

SOLUTION

(1) Write $2x$ as the First term of one binomial factor, and x as the First term of another binomial factor.

(1) $(2x\quad)(x\quad)$ with product $2x^2$

(2) As the Last terms of the two binomials, write all possible pairs of numbers whose product is $+3$, the last term of the given trinomial. Notice that, since the First terms are not both x, we must consider each pair of numbers twice.

(2) $(2x+1)(x+3)$

$(2x+3)(x+1)$

$(2x-1)(x-3)$

$(2x-3)(x-1)$

(3) Combine the Outer and Inner products of each possibility, and choose the combination that gives $+5x$, the middle term of the given trinomial.

(3) $(2x+1)(x+3)$: outer $+6x$, inner $+x$: $+6x + x = +7x$

$(2x+3)(x+1)$: outer $+2x$, inner $+3x$: $+2x + 3x = \boxed{+5x}$

$(2x-1)(x-3)$: outer $-6x$, inner $-x$: $-6x - x = -7x$

$(2x-3)(x-1)$: outer $-2x$, inner $-3x$: $-2x - 3x = -5x$

$$2x^2 + 5x + 3 = (2x+3)(x+1) \quad \text{Answer}$$

In the preceding sample problems we noted certain relationships between the signs in the trinomials and the signs in the binomial factors:

Both Positive

$2x^2 + 5x \oplus 3$
$= (2x+3)(x+1)$

Both Negative

$x^2 - 5x \oplus 6$
$= (x-2)(x-3)$

Different

$x^2 + 4x \ominus 5$
$= (x-1)(x+5)$

> When factoring trinomials of the form $ax^2 + bx + c$, in which a is positive, look at the sign of the last term, c.
>
> (1) If the sign of c is *positive*, then the signs of both numbers in the binomial factors will be the *same* as bx, the middle term of the trinomial.
>
> (2) If the sign of c is *negative*, then the signs of the numbers in the binomial factors will be *different*, one positive and one negative.

PRACTICE PROBLEMS

Factor the following trinomials.

(1) $x^2 + 8x + 15$
(2) $x^2 - 8x + 12$
(3) $x^2 - 4x - 12$
(4) $x^2 + 3x - 10$
(5) $x^2 - 9x + 14$
(6) $3x^2 + 7x + 2$
(7) $2x^2 - 5x + 2$
(8) $2x^2 + x - 3$

Answers

(1) $(x+3)(x+5)$
(2) $(x-2)(x-6)$
(3) $(x+2)(x-6)$
(4) $(x+5)(x-2)$
(5) $(x-2)(x-7)$
(6) $(3x+1)(x+2)$
(7) $(2x-1)(x-2)$
(8) $(2x+3)(x-1)$

6.4 FACTORING THE DIFFERENCE OF TWO SQUARES $A^2 - B^2$

When using the FOIL method to multiply the sum of two terms, $(A+B)$, by the difference of the same two terms, $(A-B)$, the Outer and Inner products always cancel each other, leaving a binomial of the general form $A^2 - B^2$.

$\overset{\longleftarrow -8x \longrightarrow}{(x+8)\ (x-8)}$ $\overset{\text{F}\quad\text{O}\quad\text{I}\quad\text{L}}{= x^2 - 8x + 8x - 64}$ $\underset{\longleftarrow +8x \longrightarrow}{}$ $= x^2 - 64$
$\overset{\longleftarrow -15xy \longrightarrow}{(5x+3y)(5x-3y)} = 25x^2 - 15xy + 15xy - 9y^2$ $\underset{\longleftarrow +15xy \longrightarrow}{}$ $= 25x^2 - 9y^2$
$\overset{\longleftarrow -21x \longrightarrow}{(3x+7)\ (3x-7)} = 9x^2 - 21x + 21x - 49$ $\underset{\longleftarrow +21x \longrightarrow}{}$ $= 9x^2 - 49$

Therefore, to factor a binomial of the form $A^2 - B^2$, called the **difference of two squares**, again reverse the FOIL procedure and write the binomial as the product of the two factors $(A+B)$ and $(A-B)$. That is,

$$A^2 - B^2 = (A+B)(A-B)$$

In general, to find the two factors, $(A+B)$ and $(A-B)$, use the procedure given below.

To Factor the Difference of Two Squares: $A^2 - B^2$

(1) Take the square root of the first term of the given expression, and write the result as the First term of two binomial factors.

(2) Take the square root of the second term of the given expression, and write the result as the Last term of the two binomial factors.

(3) In one of the binomial factors, join the terms with a + sign, and in the other binomial factor, join the terms with a − sign.

Factor: $x^2 - 9$

$\sqrt{x^2} = x$
$(x\quad)(x\quad)$

$\sqrt{9} = 3$
$(x\quad 3)(x\quad 3)$

$(x+3)(x-3)$
$x^2 - 9 = (x+3)(x-3)$

SAMPLE PROBLEM

Problem 45: Factor $49x^4 - 81y^2$.

PROCEDURE

(1) Take the square roots of the two given terms, and write the results as the First and Last terms of two binomial factors. In one of the binomials, join the terms with a + sign, and in the other binomial, join the terms with a − sign.

SOLUTION

(1) $\sqrt{49x^4} = 7x^2$

$\sqrt{81y^2} = 9y$

$49x^4 - 81y^2 = (7x^2 + 9y)(7x^2 - 9y)$ Answer

PRACTICE PROBLEMS

Factor the following binomials.

(1) $x^2 - 64y^2$
(2) $4x^2 - 9y^2$
(3) $a^2x^2 - 25b^2y^2$
(4) $49x^4 - y^2$

Answers
(1) $(x + 8y)(x - 8y)$
(2) $(2x + 3y)(2x - 3y)$
(3) $(ax + 5by)(ax - 5by)$
(4) $(7x^2 + y)(7x^2 - y)$

6.5 FACTORING EXPRESSIONS COMPLETELY

The three types of expressions we have factored are summarized in the table below.

Polynomials Having a Common Monomial Factor
$AB + AC + \cdots + AR = A(B + C + \cdots + R)$
Trinomials of the Form: $ax^2 + bx + c$
$ax^2 + bx + c = (mx + n)(rx + s)$
where $a, b, c, m, n, r,$ and s are specific signed numbers.
The Difference of Two Squares: $A^2 - B^2$
$A^2 - B^2 = (A + B)(A - B)$

Sometimes, after an expression has been factored, one or more of its resulting polynomial factors can be factored further. This is illustrated in the example below. Notice that after factoring out a common monomial factor of $5x$, the resulting polynomial factor, $x^2 - 16$, is the difference of two squares, which can be factored further.

$$5x^3 - 80x = 5x(x^2 - 16)$$
$$= 5x(x + 4)(x - 4)$$

When an expression cannot be factored further, like the final result above, it is said to be **factored completely**.

58 / Algebra Simplified and Self-Taught

SAMPLE PROBLEMS

Problem 46: Factor $x^4y + 2x^3y - 15x^2y$ completely.

PROCEDURE	SOLUTION
(1) Factor out the common monomial factor, x^2y.	(1) $x^4y + 2x^3y - 15x^2y$ $= x^2y(x^2 + 2x - 15)$
(2) Factor the trinomial in the parentheses.	(2) $= x^2y(x+5)(x-3)$ Answer

Problem 47: Factor $x^4 - y^4$ completely.

PROCEDURE	SOLUTION
(1) Factor the difference of two squares.	(1) $x^4 - y^4$ $= (x^2 + y^2)(x^2 - y^2)$
(2) Factor the expression in the right parentheses, which is also the difference of two squares.	(2) $= (x^2 + y^2)(x+y)(x-y)$ Answer

PRACTICE PROBLEMS

Factor the following expressions completely.

(1) $3x^2y + 9xy + 6y$
(2) $x^3 - 16x$
(3) $5x^3 + 5x^2 - 30x$
(4) $50x^2 - 8y^2$

Answers
(1) $3y(x+1)(x+2)$
(2) $x(x+4)(x-4)$
(3) $5x(x-2)(x+3)$
(4) $2(5x+2y)(5x-2y)$

6.6 SOLVING QUADRATIC EQUATIONS BY FACTORING

Equations such as $x^2 = 2x + 3$ and $5x - 6 = x^2$, in which the highest power of the variable is *two*, are called **second degree**, or **quadratic equations**.

When a quadratic equation is written in a form such that one side of the equation is 0, and the other side of the equation has the variable terms arranged in descending order of their powers, the equation is said to be written in **standard form**.

Standard Form
$ax^2 + bx + c = 0$

There are several methods of solving quadratic equations. Our method is based on the following principle:

Zero Product Principle

If the product of two factors is equal to 0, then at least one of the factors is equal to 0.

If $A \cdot B = 0$, then $A = 0$, $B = 0$, or A and $B = 0$.

To demonstrate the method, let us solve the quadratic equation $x^2 + 2 = 3x$. First, by using inverse operations, rewrite the equation in standard form. The purpose of this step is to get 0 on one side of the equation. Thus,

$$\begin{array}{rcl} x^2 + 2 &=& 3x \\ -3x & & -3x \\ \hline x^2 - 3x + 2 &=& 0 \end{array}$$

Then, by factoring the expression on the left side of the equation, transform the equation into a product of factors equal to 0. That is,

$$x^2 - 3x + 2 = 0$$
$$(x-1)(x-2) = 0$$

Finally, by using the zero product principle stated above, set each factor equal to 0, and solve for x. Since the values of x that make the factors equal to 0, also make the product equal to 0, these values will also be the roots of the original quadratic equation. Therefore,

$$x^2 - 3x + 2 = 0$$
$$(x-1)(x-2) = 0$$

$$\begin{array}{rcl|rcl} x - 1 &=& 0 & x - 2 &=& 0 \\ +1 & & +1 & +2 & & +2 \\ \hline x &=& 1 & x &=& 2 \end{array}$$

$$x = 1, \, x = 2$$

To check the roots obtained, substitute each one separately into the original equation:

Check

$$\begin{array}{c|c} x = 1 & x = 2 \\ x^2 + 2 = 3x & x^2 + 2 = 3x \\ (1)^2 + 2 \stackrel{?}{=} 3(1) & (2)^2 + 2 \stackrel{?}{=} 3(2) \\ 1 + 2 \stackrel{?}{=} 3 & 4 + 2 \stackrel{?}{=} 6 \\ 3 \stackrel{\checkmark}{=} 3 & 6 \stackrel{\checkmark}{=} 6 \end{array}$$

As this example demonstrates, most quadratic equations have two distinct roots. There are, however, some quadratic equations which have only one root. For example, consider the equation $x^2-6x+9=0$. When we factor this equation, we get $(x-3)(x-3)=0$. Since both factors are the same, they both lead to the same root, $x=3$.

The method just demonstrated for solving quadratic equations by factoring is summarized below. Not all quadratic equations are factorable, and thus other methods must be used. These methods, however, will not be discussed here.

To Solve a Quadratic Equation by Factoring:

(1) Rewrite the equation in standard form:

$$ax^2+bx+c=0$$

(2) Factor the expression on the left side of the equation.
(3) Set each factor equal to 0, and solve the resulting linear equations.

SAMPLE PROBLEMS

Problem 48: Solve $3x^2-12=x(1+2x)$, and check the roots.

PROCEDURE	SOLUTION
(1) Using the distributive law, eliminate the parentheses. Rewrite the equation in standard form.	(1) $3x^2-12=x(1+2x)$ $\begin{array}{r} 3x^2-12 = x+2x^2 \\ -2x^2 -2x^2 \\ \hline x^2-12 = x \\ -x -x \\ \hline x^2-x-12 = 0 \end{array}$
(2) Factor the expression on the left side of the equation.	(2) $x^2-x-12=0$ $(x-4)(x+3)=0$
(3) Set each factor equal to 0, and solve the resulting linear equations for x.	(3) $\begin{array}{r} x-4= 0 \\ +4 \ +4 \\ \hline x \ = 4 \end{array}$ $\quad$ $\begin{array}{r} x+3= 0 \\ -3 \ -3 \\ \hline x \ =-3 \end{array}$ $x=4, x=-3$ $\qquad$ *Answer*

(4)	Check each root separately by substituting it into the original equation.	(4)	Check $x=4$ $$3x^2-12=x(1+2x)$$ $$3(4)^2-12 \stackrel{?}{=} 4(1+2(4))$$ $$3(16)-12 \stackrel{?}{=} 4(1+8)$$ $$48-12 \stackrel{?}{=} 4(9)$$ $$36 \stackrel{\checkmark}{=} 36$$ Check $x=-3$ $$3x^2-12=x(1+2x)$$ $$3(-3)^2-12 \stackrel{?}{=} -3(2(-3)+1)$$ $$3(9)-12 \stackrel{?}{=} -3(-6+1)$$ $$27-12 \stackrel{?}{=} -3(-5)$$ $$15 \stackrel{\checkmark}{=} 15$$

Problem 49: Solve $2x^2+3x=9x$, and check the roots.

	PROCEDURE		SOLUTION
(1)	Rewrite the equation in standard form.	(1)	$$\begin{array}{r}2x^2+3x= 9x \\ -9x -9x \\ \hline 2x^2-6x= 0 \end{array}$$
(2)	Factor out the common factor $2x$.	(2)	$2x(x-3)=0$
(3)	Set each factor equal to 0, and solve the resulting linear equations for x.	(3)	$$\begin{array}{c|c} 2x=0 & x-3=0 \\ \dfrac{2x}{2}=\dfrac{0}{2} & \begin{array}{r}+3 +3 \\ \hline x = 3 \end{array} \\ x=0 & \end{array}$$ $x=0,\ x=3$ Answer
(4)	Check each root separately by substituting it into the original equation.	(4)	Check $x=0$ $$2x^2+3x=9x$$ $$2(0)^2+3(0) \stackrel{?}{=} 9(0)$$ $$0 \stackrel{\checkmark}{=} 0$$ Check $x=3$ $$2x^2+3x=9x$$ $$2(3)^2+3(3) \stackrel{?}{=} 9(3)$$ $$2(9)+9 \stackrel{?}{=} 27$$ $$27 \stackrel{\checkmark}{=} 27$$

Problem 50: Solve $5x^2 = 36 + x^2$, and check the roots.

PROCEDURE	SOLUTION
(1) Rewrite the equation in standard form.	(1) $\begin{aligned} 5x^2 &= 36 + x^2 \\ -x^2 & \quad\quad -x^2 \\ \hline 4x^2 &= 36 \\ -36 & \quad -36 \\ \hline 4x^2 - 36 &= 0 \end{aligned}$
(2) Factor the difference of two squares.	(2) $4x^2 - 36 = 0$ $(2x+6)(2x-6) = 0$
(3) Set each factor equal to 0 and solve the resulting linear equations for x.	(3) $\begin{aligned} 2x+6 &= 0 & 2x-6 &= 0 \\ -6 & \; -6 & +6 & \; +6 \\ \hline 2x &= -6 & 2x &= 6 \\ \frac{2x}{2} &= \frac{-6}{2} & \frac{2x}{2} &= \frac{6}{2} \\ x &= -3 & x &= 3 \end{aligned}$ $x = -3, x = 3$ Answer
(4) Check each root separately by substituting it into the original equation.	(4) Check $x = -3$ $5x^2 = 36 + x^2$ $5(-3)^2 \stackrel{?}{=} 36 + (-3)^2$ $5(9) \stackrel{?}{=} 36 + 9$ $45 \stackrel{\checkmark}{=} 45$ Check $x = 3$ $5x^2 = 36 + x^2$ $5(3)^2 \stackrel{?}{=} 36 + (3)^2$ $5(9) \stackrel{?}{=} 36 + 9$ $45 \stackrel{\checkmark}{=} 45$

PRACTICE PROBLEMS

Solve each of the following quadratic equations for x.

(1) $x^2 - 7x + 12 = 0$
(2) $x^2 - 18x + 81 = 0$
(3) $x^2 - 18 = 3x$
(4) $x^2 + x = 30$
(5) $3x^2 + 12x = 0$
(6) $2x^2 - 5x = x$
(7) $2x^2 - 50 = 0$
(8) $4x^2 = x^2 + 48$

Answers

(1) 3, 4
(2) 9
(3) 6, −3
(4) −6, 5
(5) 0, −4
(6) 0, 3
(7) 5, −5
(8) 4, −4

6.7 SOLVING INCOMPLETE QUADRATIC EQUATIONS

$ax^2 + c = 0$

Quadratic equations such as $x^2 - 9 = 0$, which when written in standard form do not contain a middle term, bx, are called **incomplete quadratic equations**.

One way of solving incomplete quadratic equations is by the factoring method described in the last section. Note that in general, the expression on the left side of the equation will be the difference of two squares. For example,

$$x^2 - 9 = 0$$
$$(x+3)(x-3) = 0$$

$$\begin{array}{c|c} x+3 = 0 & x-3 = 0 \\ -3 \quad -3 & +3 \quad +3 \\ \hline x = -3 & x = 3 \end{array}$$

$$x = -3, x = 3$$

Another way of solving incomplete quadratic equations is to rewrite the equation in the form $x^2 = n$, where n is some positive number, and then take the square root of both sides. That is,

$$\begin{array}{r} x^2 - 9 = 0 \\ +9 \quad +9 \\ \hline x^2 = 9 \end{array}$$
$$x^2 = 9$$

$\sqrt{x^2} = +\sqrt{9}$ or $\sqrt{x^2} = -\sqrt{9}$
$x = +3$ or $x = -3$

Notice that we must consider both $+\sqrt{9}$ and $-\sqrt{9}$, since both of these numbers squared equal 9. This second procedure is summarized below.

To Solve an Incomplete Quadratic Equation:

(1) Rewrite the equation in the form $x^2 = n$, where n is some positive number.
(2) Take the square root of n. The two roots are $x = +\sqrt{n}$ or $x = -\sqrt{n}$, which is sometimes denoted $x = \pm\sqrt{n}$.

SAMPLE PROBLEM

Problem 51: Solve $5x^2 = 3x^2 + 32$, and check the roots.

PROCEDURE

(1) Rewrite the equation in the form $x^2 = n$.

SOLUTION

(1)
$$\begin{array}{r} 5x^2 = 3x^2 + 32 \\ -3x^2 \quad -3x^2 \\ \hline 2x^2 = 32 \end{array}$$
$$\frac{\cancel{2}x^2}{\cancel{2}} = \frac{32}{2}$$
$$x^2 = 16$$

(2) Take the positive and negative square roots of 16.	(2) $x = +\sqrt{16}$ and $x = -\sqrt{16}$ $x = +4, x = -4$ Answer
(3) Check each root separately by substituting it into the original equation.	(3) Check $x = 4$ $5x^2 = 3x^2 + 32$ $5(4)^2 \stackrel{?}{=} 3(4)^2 + 32$ $5(16) \stackrel{?}{=} 3(16) + 32$ $80 \stackrel{\checkmark}{=} 80$ Check $x = -4$ $5x^2 = 3x^2 + 32$ $5(-4)^2 \stackrel{?}{=} 3(-4)^2 + 32$ $5(16) = 3(16) + 32$ $80 \stackrel{\checkmark}{=} 80$

PRACTICE PROBLEMS

Solve each of the following incomplete quadratic equations for x.

(1) $x^2 - 49 = 0$
(2) $4x^2 = 100$
(3) $3x^2 - 32 = x^2$
(4) $2x^2 + 18 = 3x^2 - 7$

Answers	
(1) $+7, -7$	(2) $+5, -5$
(3) $+4, -4$	(4) $+5, -5$

7

7.1 SIMPLIFYING SQUARE ROOTS OF NUMBERS

As you recall, numbers such as 4, 9, and 16, whose square roots are whole numbers, are called **perfect squares**. To **simplify** a square root means to remove any perfect square factors from under the square root sign. The method we use to do this is based on the following principle.

$\sqrt{4 \cdot 9} = \sqrt{4} \cdot \sqrt{9}$

$\sqrt{36} = 2 \cdot 3$

$6 = 6$

> *Multiplication Principle of Square Roots*
>
> The square root of a product is equal to the product of the square roots.
>
> $$\sqrt{a \cdot b} = \sqrt{a} \cdot \sqrt{b}$$

To demonstrate the method, let us simplify $\sqrt{18}$. First, by writing $18 = 9 \cdot 2$, transform the number under the square root sign into a product containing a perfect square factor, 9:

$$\sqrt{18} = \sqrt{9 \cdot 2}$$

Then, by using the multiplication principle just stated, write the square root of the product as the product of the square roots.

$$\sqrt{18} = \sqrt{9 \cdot 2}$$
$$= \sqrt{9} \cdot \sqrt{2}$$

Finally, take the square root of the perfect square factor and get

$$\sqrt{18} = \sqrt{9 \cdot 2}$$
$$= \sqrt{9} \cdot \sqrt{2}$$
$$= 3\sqrt{2}$$

$\sqrt{720} = \sqrt{36 \cdot 20}$

$= \sqrt{36} \cdot \sqrt{20}$

$= 6\sqrt{20}$

$= 6\sqrt{4 \cdot 5}$

$= 6\sqrt{4} \cdot \sqrt{5}$

$= 6 \cdot 2\sqrt{5}$

$= 12\sqrt{5}$

> *To Simplify Square Roots*:
>
> (1) Express the number under the square root sign as the product of two factors, one of which is a perfect square.
> (2) Rewrite the square root of the product as the product of the square roots.
> (3) Take the square root of the perfect square factor.
> (4) Repeat the process until all the perfect square factors have been removed from under the square root sign.

Note that this procedure depends on the number being expressed as a *product*. As the examples below illustrate, the same procedure *cannot* be used if the number is expressed as a *sum* or a *difference*. That is, $\sqrt{a+b} \neq \sqrt{a} + \sqrt{b}$, and $\sqrt{a-b} \neq \sqrt{a} - \sqrt{b}$.

Square Root of Sum	*Square Root of Difference*
$\sqrt{9+16} \stackrel{?}{=} \sqrt{9} + \sqrt{16}$	$\sqrt{100-36} \stackrel{?}{=} \sqrt{100} - \sqrt{36}$
$\sqrt{25} \stackrel{?}{=} 3+4$	$\sqrt{64} \stackrel{?}{=} 10-6$
$5 \neq 7$	$8 \neq 4$

PRACTICE PROBLEMS

Simplify the following square roots.

(1) $\sqrt{75}$ (2) $\sqrt{128}$

(3) $\sqrt{40}$ (4) $\sqrt{63}$

Answers			
(1)	$5\sqrt{3}$	(2)	$8\sqrt{2}$
(3)	$2\sqrt{10}$	(4)	$3\sqrt{7}$

7.2 SIMPLIFYING SQUARE ROOTS OF VARIABLES

To simplify square roots containing variable factors, use the following general principle:

> The square root of a variable raised to an *even power* is the variable raised to one-half that even power.
>
> $$\sqrt{x^m} = x^{\frac{1}{2}m}$$

$\sqrt{x^2} = x$
$\sqrt{x^6} = x^3$
$\sqrt{x^{10}} = x^5$

If the variable is raised to an *odd power*, rewrite it as the product of two factors, one with a power of 1 less than the given odd power, and the other with a power of 1. For example,

$$\sqrt{x^7} = \sqrt{x^6 \cdot x^1}$$
$$= \sqrt{x^6} \cdot \sqrt{x^1}$$
$$= x^3 \sqrt{x}$$

SAMPLE PROBLEM

Problem 52: Simplify $\sqrt{75x^2y^7z^3}$.

PROCEDURE		SOLUTION	
(1) Express 75 as the product of the perfect square factor 25 and the factor 3.	(1)	$\sqrt{75x^2y^7z^3}$ $= \sqrt{25 \cdot 3x^2y^7z^3}$	
(2) Rewrite y^7 as $y^6 \cdot y$, and z^3 as $z^2 \cdot z$.	(2)	$= \sqrt{25 \cdot 3 \cdot x^2 \cdot y^6 \cdot y \cdot z^2 \cdot z}$	
(3) Group all the perfect square factors together.	(3)	$= \sqrt{25x^2y^6z^2 \cdot 3yz}$	
(4) Use the multiplication principle to separate the two groups.	(4)	$= \sqrt{25x^2y^6z^2} \cdot \sqrt{3yz}$	
(5) Take the square root of the perfect square factors.	(5)	$= 5xy^3z\sqrt{3yz}$	Answer

PRACTICE PROBLEMS

Answers
(1) $5xy^2\sqrt{2}$
(2) $4x^2\sqrt{5x}$
(3) $10x^3y^3\sqrt{7y}$
(4) $2xy\sqrt{3xy}$

Simplify the following square roots.

(1) $\sqrt{50x^2y^4}$ (2) $\sqrt{80x^5}$

(3) $\sqrt{700x^6y^7}$ (4) $\sqrt{12x^3y^3}$

7.3 ADDING AND SUBTRACTING SQUARE ROOTS

Like Square Roots

$-2\sqrt{ax}$, $\sqrt{ax}$, $3\sqrt{ax}$

Square roots which have identical expressions under the square root signs are called **like square roots**. To add and subtract like square roots, simply add and subtract (combine) their coefficients. Note how this is similar to adding and subtracting like algebraic terms:

$2\sqrt{7} + 4\sqrt{7} - 9\sqrt{7}$
$= (2+4-9)\sqrt{7}$
$= -3\sqrt{7}$

$3\sqrt{xy} - \sqrt{xy} + 2\sqrt{xy}$
$= (3-1+2)\sqrt{xy}$
$= 4\sqrt{xy}$

68 / *Algebra Simplified and Self-Taught*

In general, square roots which do not have identical expressions under their square root signs (unlike square roots) cannot be combined into a single term. For example, expressions such as $\sqrt{5}+\sqrt{3}$ and $\sqrt{x}-\sqrt{y}$ cannot be simplified further. In some cases, however, it may be possible to change unlike square roots into like square roots by simplifying them (removing perfect square factors):

$$\sqrt{20}+6\sqrt{5}$$
$$=\sqrt{4\cdot 5}+6\sqrt{5}$$
$$=\sqrt{4}\cdot\sqrt{5}+6\sqrt{5}$$
$$=2\sqrt{5}+6\sqrt{5}$$
$$=8\sqrt{5}$$

SAMPLE PROBLEM

Problem 53: Simplify and combine $\sqrt{2x^3}+3x\sqrt{8x}-x\sqrt{2x}$.

PROCEDURE	SOLUTION
(1) Using the multiplication principle, remove all the perfect square factors from under the square root signs.	(1) $\sqrt{2x^3}+3x\sqrt{8x}-x\sqrt{2x}$ $=\sqrt{x^2\cdot 2x}+3x\sqrt{4\cdot 2x}-x\sqrt{2x}$ $=\sqrt{x^2}\cdot\sqrt{2x}+3x\sqrt{4}\cdot\sqrt{2x}-x\sqrt{2x}$ $=x\sqrt{2x}+3x\cdot 2\sqrt{2x}-x\sqrt{2x}$ $=x\sqrt{2x}+6x\sqrt{2x}-x\sqrt{2x}$
(2) Combine the coefficients.	(2) $=(x+6x-x)\sqrt{2x}$ $=6x\sqrt{2x}$ *Answer*

PRACTICE PROBLEMS

Perform the following additions and subtractions.

(1) $5\sqrt{2x}-9\sqrt{2x}+\sqrt{2x}$ (2) $\sqrt{45}+7\sqrt{125}$

(3) $\sqrt{12}-6\sqrt{3}+2\sqrt{48}$ (4) $6\sqrt{3x^5}-x^2\sqrt{12x}$

Answers

(1) $-3\sqrt{2x}$ (2) $38\sqrt{5}$

(3) $4\sqrt{3}$ (4) $4x^2\sqrt{3x}$

7.4 MULTIPLYING SQUARE ROOTS

The multiplication principle used to simplify square roots can be stated in reverse to multiply square roots.

$\sqrt{4} \cdot \sqrt{9} = \sqrt{4 \cdot 9}$

$2 \cdot 3 = \sqrt{36}$

$6 = 6$

> A product of square roots is equal to the square root of the product.
> $$\sqrt{a} \cdot \sqrt{b} = \sqrt{a \cdot b}$$

Consider the following examples.

$$\sqrt{18} \cdot \sqrt{2} = \sqrt{18 \cdot 2}$$
$$= \sqrt{36}$$
$$= 6$$

$$\sqrt{2xy} \cdot \sqrt{32xy} = \sqrt{2xy \cdot 32xy}$$
$$= \sqrt{64x^2 y^2}$$
$$= 8xy$$

Remember that when square roots have coefficients other than 1, the square roots and coefficients must be multiplied separately:

$$(3\sqrt{5})(7\sqrt{2}) = (3 \cdot 7)(\sqrt{5} \cdot \sqrt{2})$$
$$= 21\sqrt{5 \cdot 2}$$
$$= 21\sqrt{10}$$

A common mistake that is made at this point is to continue multiplying and write $21\sqrt{10} = \sqrt{210}$. This is an error, however, because the number outside the square root *cannot* be multiplied by the number underneath the square root.

SAMPLE PROBLEM

Problem 54: Multiply $(2\sqrt{6xy})(5\sqrt{3x^3 y})(3\sqrt{y})$.

PROCEDURE	SOLUTION
(1) Multiply the numerical coefficients and square roots separately.	(1) $(2\sqrt{6xy})(5\sqrt{3x^3 y})(3\sqrt{y})$ $= 2 \cdot 5 \cdot 3 \sqrt{(6xy)(3x^3 y)(y)}$ $= 30\sqrt{18 x^4 y^3}$
(2) Simplify the square root by removing perfect square factors.	(2) $= 30\sqrt{9 \cdot 2 \cdot x^4 \cdot y^2 \cdot y}$ $= 30\sqrt{9 x^4 y^2 \cdot 2y}$ $= 30\sqrt{9 x^4 y^2} \cdot \sqrt{2y}$ $= 30 \cdot 3 x^2 y \sqrt{2y}$ $= 90 x^2 y \sqrt{2y}$ *Answer*

PRACTICE PROBLEMS

Perform the following multiplications. Simplify all answers.

(1) $(3\sqrt{32})(5\sqrt{2})$
(2) $(-2\sqrt{3})(6\sqrt{27})$
(3) $\sqrt{18x^3} \cdot \sqrt{2x}$
(4) $\sqrt{3x^3y} \cdot \sqrt{15xy}$

Answers
(1) 120
(2) −108
(3) $6x^2$
(4) $3x^2y\sqrt{5}$

7.5 DIVIDING SQUARE ROOTS

To divide square roots, we use a principle similar to the one used in the last section for multiplying square roots.

> The quotient of two square roots is equal to the square root of the quotient.
>
> $$\frac{\sqrt{a}}{\sqrt{b}} = \sqrt{\frac{a}{b}}$$

$$\frac{\sqrt{100}}{\sqrt{4}} = \sqrt{\frac{100}{4}}$$

$$\frac{10}{2} = \sqrt{25}$$

$$5 = 5$$

Consider these examples.

$$\frac{\sqrt{45}}{\sqrt{5}} = \sqrt{\frac{45}{5}} \qquad \frac{\sqrt{48x^3}}{\sqrt{3x}} = \sqrt{\frac{48x^3}{3x}}$$

$$= \sqrt{9} \qquad\qquad = \sqrt{16x^2}$$

$$= 3 \qquad\qquad = 4x$$

As before, remember that when square roots have coefficients other than 1, the square roots and coefficients must be divided separately:

$$\frac{12\sqrt{75}}{2\sqrt{3}} = \frac{12}{2}\sqrt{\frac{75}{3}}$$

$$= 6\sqrt{25}$$

$$= 6 \cdot 5$$

$$= 30$$

SAMPLE PROBLEM

Problem 55: Divide $\dfrac{6\sqrt{56xy^4}}{2\sqrt{7x^5y}}$.

PROCEDURE

(1) Divide the numerical coefficients and square roots separately.

(2) Simplify the square root by removing perfect square factors.

SOLUTION

(1)
$$\dfrac{6\sqrt{56xy^4}}{2\sqrt{7x^5y}}$$

$$= \dfrac{6}{2}\sqrt{\dfrac{56xy^4}{7x^5y}}$$

$$= 3\sqrt{\dfrac{8y^3}{x^4}}$$

(2)
$$= 3\sqrt{\dfrac{4y^2 \cdot 2y}{x^4}}$$

$$= 3\sqrt{\dfrac{4y^2}{x^4}} \cdot \sqrt{2y}$$

$$= 3 \cdot \dfrac{2y}{x^2}\sqrt{2y}$$

$$= \dfrac{6y}{x^2}\sqrt{2y} \qquad \text{Answer}$$

PRACTICE PROBLEMS

Perform the following divisions. Simplify all answers.

(1) $\dfrac{6\sqrt{75}}{2\sqrt{3}}$

(2) $\dfrac{\sqrt{98x^3}}{\sqrt{2x}}$

(3) $\dfrac{7\sqrt{54x^5}}{\sqrt{2x^2}}$

(4) $\dfrac{15\sqrt{8x^5y^5}}{5\sqrt{2xy^3}}$

Answers

(1) 15   (2) $7x$

(3) $21x\sqrt{3x}$ (4) $6x^2y$

8

8.1 ALGEBRAIC FRACTIONS

Fractions which contain variables, such as $\frac{x}{4}$, $\frac{3x}{2y}$, and $\frac{7}{3x-4}$, are called **algebraic fractions**. Any algebraic expression can be put into the form of an algebraic fraction by placing it over a denominator of 1. For example, $3x = \frac{3x}{1}$, and $2x^3 - 5 = \frac{2x^3 - 5}{1}$.

Since division by 0 is undefined, we must be careful not to replace any variables in the denominator of an algebraic fraction by numbers which would make its value 0. For example, the fraction $\frac{5x}{x-2}$ has no meaning for $x = 2$.

$$\frac{5x}{x-2} = \frac{5 \cdot 2}{2-2}$$

$$= \frac{10}{0} ??$$

8.2 REDUCING ALGEBRAIC FRACTIONS TO LOWEST TERMS

We know from arithmetic that to reduce a fraction to lowest terms means to divide the numerator and denominator by common factors until no common factors remain. As shown opposite, this is equivalent to expressing the numerator and denominator as a product, and then cancelling the common factors.

When both the numerator and denominator of an algebraic fraction are monomials, we can cancel common factors by using the rules shown opposite for dividing powers of the same base. For example,

$$\frac{10}{12} = \frac{5 \cdot \cancel{2}}{6 \cdot \cancel{2}} = \frac{5}{6}$$

$$\frac{x^m}{x^n} = \begin{cases} x^{m-n}, & m > n \\ 1, & m = n \\ \frac{1}{x^{n-m}}, & m < n \end{cases}$$

$$\frac{6x^3y^2}{8xy^3} = \frac{\overset{3}{\cancel{6}} \cdot \cancel{x} \cdot x \cdot x \cdot \cancel{y} \cdot \cancel{y}}{\underset{4}{\cancel{8}} \cdot \cancel{x} \cdot y \cdot \cancel{y} \cdot \cancel{y}}$$

$$= \frac{3x^{3-1}}{4y^{3-2}}$$

$$= \frac{3x^2}{4y}$$

When the expressions in the numerator or denominator are polynomials (more than one term), we cannot cancel common factors until the expressions are rewritten in factored form:

$$\frac{3x^2+6x}{12x} = \frac{\overset{1}{\cancel{3x}}(x+2)}{\underset{4}{\cancel{12x}}}$$

$$= \frac{x+2}{4}$$

$$\frac{x^2-9}{5x+15} = \frac{(x+3)(x-3)}{5(x+3)}$$

$$= \frac{x-3}{5}$$

A common mistake in reducing fractions containing polynomials is to reduce common terms before the expressions are rewritten in factored form. Some typical errors are shown below.

Errors in Cancelling

$$\frac{x+3}{y+3} = \frac{x+\cancel{3}}{y+\cancel{3}}$$

$$= \frac{x}{y} \quad \text{False}$$

$$\frac{x^2+6}{x+3} = \frac{\overset{x}{\cancel{x^2}}+\overset{3}{\cancel{6}}}{\cancel{x}+\cancel{2}}$$

$$= x+3 \quad \text{False}$$

Correct

$$\frac{a \cdot c}{b \cdot c} = \frac{a \cdot \cancel{c}}{b \cdot \cancel{c}} = \frac{a}{b}$$

Incorrect

$$\frac{a+c}{b+c} = \frac{a+\cancel{c}}{b+\cancel{c}} = \frac{a}{b}$$

Remember that the only correct way of reducing fractions by cancelling is to cancel common *factors*, *not* common *terms*.

SAMPLE PROBLEM

Problem 56: Reduce the fraction $\dfrac{x^2-3x-10}{x^2-25}$.

PROCEDURE	SOLUTION
(1) Factor the expressions in the numerator and denominator.	(1) $\dfrac{x^2-3x-10}{x^2-25}$ $= \dfrac{(x+2)(x-5)}{(x+5)(x-5)}$
(2) Cancel the common factor, $(x-5)$.	(2) $= \dfrac{(x+2)\cancel{(x-5)}}{(x+5)\cancel{(x-5)}}$ $= \dfrac{x+2}{x+5}$ Answer

PRACTICE PROBLEMS

Reduce each of the following fractions to lowest terms.

(1) $\dfrac{4x^3y^2}{8xy^5}$

(2) $\dfrac{6x^3+2x}{4x}$

(3) $\dfrac{x^2+7x+12}{2x+6}$

(4) $\dfrac{x^2-64}{3x-24}$

Answers

(1) $\dfrac{x^2}{2y^3}$

(2) $\dfrac{3x^2+1}{2}$

(3) $\dfrac{x+4}{2}$

(4) $\dfrac{x+8}{3}$

8.3 ADDING AND SUBTRACTING LIKE ALGEBRAIC FRACTIONS

To add or subtract algebraic fractions having the same denominator (*like* algebraic fractions), add or subtract their numerators, and place the result over the common denominator. In each case, reduce the resulting fraction to lowest terms:

$$\frac{3x^2}{8y} + \frac{7x^2}{8y} - \frac{4x^2}{8y} = \frac{3x^2+7x^2-4x^2}{8y}$$

$$= \frac{6x^2}{8y}$$

$$= \frac{3x^2}{4y}$$

$$\frac{a}{c} + \frac{b}{c} = \frac{a+b}{c}$$

$$\frac{a}{c} - \frac{b}{c} = \frac{a-b}{c}$$

SAMPLE PROBLEMS

Problem 57: Add $\dfrac{x+1}{4x+6} + \dfrac{x+2}{4x+6}$.

PROCEDURE	SOLUTION
(1) Add the numerators, and place the result over the common denominator.	(1) $\dfrac{x+1}{4x+6} + \dfrac{x+2}{4x+6}$ $= \dfrac{x+1+x+2}{4x+6}$ $= \dfrac{2x+3}{4x+6}$

(2) Factor the expression in the denominator, and then cancel the common factor $(2x+3)$.

(2) $$= \frac{\cancel{2x+3}}{2\cancel{(2x+3)}}$$

$$= \frac{1}{2}$$ Answer

Problem 58: Subtract $\dfrac{5x-2}{2xy} - \dfrac{3x-7}{2xy}$.

PROCEDURE

(1) Subtract the numerators, and place the result over the common denominator. Remember to change the sign of each term being subtracted to its opposite sign.

SOLUTION

(1) $$\frac{5x-2}{2xy} - \frac{3x-7}{2xy}$$

$$= \frac{(5x-2)-(3x-7)}{2xy}$$

$$= \frac{5x-2-3x+7}{2xy}$$

$$= \frac{2x+5}{2xy}$$ Answer

PRACTICE PROBLEMS

Perform the following additions and subtractions. Reduce all answers to lowest terms.

Answers

(1) $\dfrac{3x}{a}$

(2) $\dfrac{8x-1}{y}$

(3) $\dfrac{5x+8}{ab}$

(4) 3

(1) $\dfrac{2x}{3a} + \dfrac{8x}{3a} - \dfrac{x}{3a}$

(2) $\dfrac{6x-4}{y} + \dfrac{2x+3}{y}$

(3) $\dfrac{7x+3}{ab} - \dfrac{2x-5}{ab}$

(4) $\dfrac{4x+1}{x-2} - \dfrac{x+7}{x-2}$

8.4 ADDING AND SUBTRACTING UNLIKE ALGEBRAIC FRACTIONS

To add or subtract algebraic fractions having different denominators (*unlike* algebraic fractions), first change them to equivalent fractions having a common denominator, and then proceed as in the last section. Remember that a simple way of obtaining a common denominator is to multiply the given denominators.

For example, to add $\frac{a}{b} + \frac{c}{d}$ first change the fractions to equivalent fractions having the common denominator bd (the product of the denominators). As demonstrated below, to change the fractions to this denominator, multiply $\frac{a}{b}$ by $\frac{d}{d}$ and $\frac{c}{d}$ by $\frac{b}{b}$.

$$\frac{a}{b} + \frac{c}{d} = \frac{a}{b} \cdot \frac{d}{d} + \frac{c}{d} \cdot \frac{b}{b}$$

$$= \frac{ad}{bd} + \frac{bc}{bd}$$

$$= \frac{ad + bc}{bd}$$

The following shortcut can be used to obtain the same result:

To Add or Subtract Two Unlike Fractions:

Cross multiply the numerators and denominators, and place their sum or difference over the product of the denominators (the common denominator).

$$\frac{a}{b} \times \frac{c}{d} = \frac{ad + bc}{bd} \qquad \Big| \qquad \frac{a}{b} \times \frac{c}{d} = \frac{ad - bc}{bd}$$

SAMPLE PROBLEMS

Problem 59: Add $\frac{2x}{5} + \frac{3x}{7}$.

PROCEDURE	SOLUTION
(1) Cross multiply the numerators and denominators, and place their sum over the product of the denominators.	(1) $\frac{2x}{5} + \frac{3x}{7}$ $= \frac{7(2x) + 5(3x)}{5 \cdot 7}$ $= \frac{14x + 15x}{35}$ $= \frac{29x}{35}$ Answer

Problem 60: Add $\dfrac{5x}{x+3} + \dfrac{2x}{x-3}$.

PROCEDURE	SOLUTION
(1) Cross multiply the numerators and denominators, and place their sum over the product of the denominators.	(1) $\dfrac{5x}{x+3} + \dfrac{2x}{x-3}$ $= \dfrac{5x(x-3)+2x(x+3)}{(x+3)(x-3)}$ $= \dfrac{5x^2-15x+2x^2+6x}{x^2-9}$ $= \dfrac{7x^2-9x}{x^2-9}$ Answer

Problem 61: Subtract $\dfrac{7}{xy^2} - \dfrac{3}{x^2y}$.

PROCEDURE	SOLUTION
(1) Cross multiply the numerators and denominators, and place their difference over the product of the denominators.	(1) $\dfrac{7}{xy^2} - \dfrac{3}{x^2y}$ $= \dfrac{7(x^2y)-3(xy^2)}{(xy^2)(x^2y)}$ $= \dfrac{7x^2y-3xy^2}{x^3y^3}$
(2) Factor the expression in the numerator, and cancel the common factor xy.	(2) $= \dfrac{\overset{1}{\cancel{xy}}(7x-3y)}{\underset{x^2y^2}{\cancel{x^3y^3}}}$ $= \dfrac{7x-3y}{x^2y^2}$ Answer

PRACTICE PROBLEMS

Perform the following additions and subtractions. Reduce all answers to lowest terms.

(1) $\dfrac{3x}{7} + \dfrac{2x}{5}$

(2) $\dfrac{8}{x^2} - \dfrac{5}{x}$

Answers

(1) $\dfrac{29x}{35}$

(2) $\dfrac{8-5x}{x^2}$

(3) $\dfrac{4}{x+5} + \dfrac{3}{x-5}$ (4) $\dfrac{7}{x+1} - \dfrac{3}{x+2}$

> **Answers**
> (3) $\dfrac{7x-5}{x^2-25}$
> (4) $\dfrac{4x+11}{x^2+3x+2}$

8.5 MULTIPLYING ALGEBRAIC FRACTIONS

To multiply algebraic fractions, multiply the numerators by the numerators and the denominators by the denominators:

$$\dfrac{3x}{y} \cdot \dfrac{2x^2}{5y} = \dfrac{3x \cdot 2x^2}{y \cdot 5y}$$

$$= \dfrac{6x^3}{5y^2}$$

$$\dfrac{a}{b} \cdot \dfrac{c}{d} = \dfrac{a \cdot c}{b \cdot d}$$

If the numerators and denominators contain common factors, they can be cancelled before multiplying:

$$\dfrac{12x^2}{y} \cdot \dfrac{5}{4x} = \dfrac{\overset{3x}{\cancel{12x^2}}}{y} \cdot \dfrac{5}{\underset{1}{\cancel{4x}}}$$

$$= \dfrac{3x \cdot 5}{y \cdot 1}$$

$$= \dfrac{15x}{y}$$

Remember that if the numerators or denominators are polynomials (more than one term), they must first be factored before common factors can be cancelled.

SAMPLE PROBLEM

Problem 62: Multiply $\dfrac{2x-10}{x-3} \cdot \dfrac{x^2-9}{8x-40}$.

PROCEDURE	SOLUTION
(1) Factor the expressions in the numerator and denominator.	(1) $\dfrac{2x-10}{x-3} \cdot \dfrac{x^2-9}{8x-40}$ $= \dfrac{2(x-5)}{x-3} \cdot \dfrac{(x+3)(x-3)}{8(x-5)}$

(2) Cancel the common factors.

(2) $$= \frac{\overset{1}{\cancel{2}}\overset{1}{\cancel{(x-5)}}}{\underset{1}{\cancel{x-3}}} \cdot \frac{(x+3)\overset{1}{\cancel{(x-3)}}}{\underset{4}{\cancel{8}}\underset{1}{\cancel{(x-5)}}}$$

(3) Multiply the remaining factors.

(3) $$= \frac{x+3}{4} \quad \text{Answer}$$

PRACTICE PROBLEMS

Perform the following multiplications. Reduce all answers to lowest terms.

Answers
(1) $\dfrac{12x^3}{y^4}$
(2) $\dfrac{10x^2}{y^3}$
(3) $\dfrac{2x+2}{y}$
(4) $\dfrac{3x-15}{7}$

(1) $\dfrac{4x^2}{y} \cdot \dfrac{3x}{y^3}$

(2) $\dfrac{5x^3}{2y^4} \cdot \dfrac{4y}{x}$

(3) $\dfrac{8y}{4x-4} \cdot \dfrac{x^2-1}{y^2}$

(4) $\dfrac{3x+6}{x+1} \cdot \dfrac{x^2-4x-5}{7x+14}$

8.6 DIVIDING ALGEBRAIC FRACTIONS

$$\frac{a}{b} \div \frac{c}{d} = \frac{a}{b} \cdot \frac{d}{c}$$

To divide algebraic fractions, first invert the divisor (the fraction following the division symbol), and then multiply the resulting fractions as in the last section. Remember that if the divisor is not a fraction, first place it over a denominator of 1 before inverting it:

$$\frac{2x}{3y} \div 5y = \frac{2x}{3y} \div \frac{5y}{1}$$

$$= \frac{2x}{3y} \cdot \frac{1}{5y}$$

$$= \frac{2x}{15y^2}$$

SAMPLE PROBLEM

Problem 63: Divide $\dfrac{x^2}{2x-8} \div \dfrac{x^3}{x^2-16}$.

PROCEDURE

(1) Invert the divisor, and change the division to multiplication.

(2) Factor the expressions in the numerator and denominator.

(3) Cancel the common factors.

(4) Multiply the remaining factors.

SOLUTION

(1) $\dfrac{x^2}{2x-8} \div \dfrac{x^3}{x^2-16}$

$= \dfrac{x^2}{2x-8} \cdot \dfrac{x^2-16}{x^3}$

(2) $= \dfrac{x^2}{2(x-4)} \cdot \dfrac{(x+4)(x-4)}{x^3}$

(3) $= \dfrac{\overset{1}{\cancel{x^2}}}{2\cancel{(x-4)}} \cdot \dfrac{(x+4)\cancel{(x-4)}}{\underset{x}{\cancel{x^3}}}$

(4) $= \dfrac{x+4}{2x}$ Answer

PRACTICE PROBLEMS

Perform the following divisions. Reduce all answers to lowest terms.

(1) $\dfrac{x^2}{3} \div \dfrac{x^3}{6}$

(2) $\dfrac{8x^2}{3y^3} \div \dfrac{2x}{9y}$

(3) $\dfrac{x^2-9}{x^2} \div \dfrac{2x+6}{x}$

(4) $\dfrac{x^2-5x+6}{3x} \div \dfrac{x-2}{9x}$

Answers

(1) $\dfrac{2}{x}$

(2) $\dfrac{12x}{y^2}$

(3) $\dfrac{x-3}{2x}$

(4) $3x-9$

8.7 SIMPLIFYING COMPLEX ALGEBRAIC FRACTIONS

An algebraic fraction which contains at least one other fraction in its numerator or denominator is called a **complex algebraic fraction**.

To simplify a complex algebraic fraction, multiply every term in the fraction by a common denominator of the fractions contained within it. For example, to simplify the complex fraction shown below, multi-

Complex Fractions

$\dfrac{3+\dfrac{x}{5}}{2}, \quad \dfrac{x+\dfrac{1}{x}}{y+\dfrac{1}{y}}$

ply every term by $2x$, a common denominator of the fractions $\dfrac{3}{x}$ and $\dfrac{z}{2}$. As a result, all the denominators cancel, leaving a fraction free of other fractions.

$$\frac{y+\dfrac{3}{x}}{\dfrac{z}{2}} = \frac{2x\cdot y + 2\not{x}\cdot\dfrac{3}{\not{x}}}{\not{2}x\cdot\dfrac{z}{\not{2}}}$$

$$= \frac{2xy+6}{xz}$$

SAMPLE PROBLEM

Problem 64: Simplify $\dfrac{x+\dfrac{2}{x}}{y+\dfrac{3}{y}}$.

PROCEDURE

(1) Multiply every term in the fraction by xy, a common denominator of $\dfrac{2}{x}$ and $\dfrac{3}{y}$. Cancel the denominators.

SOLUTION

(1)
$$\frac{x+\dfrac{2}{x}}{y+\dfrac{3}{y}}$$

$$= \frac{xy\cdot x + \not{x}y\cdot\dfrac{2}{\not{x}}}{xy\cdot y + x\not{y}\cdot\dfrac{3}{\not{y}}}$$

$$= \frac{x^2 y + 2y}{xy^2 + 3x} \qquad \text{Answer}$$

PRACTICE PROBLEMS

Simplify the following complex fractions.

Answers

(1) $\dfrac{bx}{ay}$

(2) $\dfrac{5x+2}{6x}$

(3) $\dfrac{2y-3x}{5xy}$

(4) $\dfrac{4x^2+3x}{7}$

(1) $\dfrac{\dfrac{x}{y}}{\dfrac{a}{b}}$

(2) $\dfrac{5+\dfrac{2}{x}}{6}$

(3) $\dfrac{\dfrac{2}{x}-\dfrac{3}{y}}{5}$

(4) $\dfrac{4+\dfrac{3}{x}}{\dfrac{7}{x^2}}$

ALGEBRA PRACTICE TEST

1. Evaluate: $5^2 + 3(7-5)^3$

   (A) 224
   (B) 28
   (C) 49
   (D) 78
   (E) 241

2. For $a=2$, $b=1$, and $x=-3$, what is the value of $\dfrac{\sqrt{b^2-4ax}}{a-x}$?

   (A) $\dfrac{3\sqrt{2}}{5}$
   (B) 5
   (C) -5
   (D) 1
   (E) $\dfrac{2\sqrt{6}}{5}$

3. Add and Subtract: $(4x^2+y^2)+(x^2-5y^2)-(2x^2-y^2)$

   (A) $3x^2-4y^2$
   (B) x^2y^2
   (C) $3x^6-3y^6$
   (D) $3x^2-3y^2$
   (E) $-8x^2-5y^2$

4. Multiply: $(-2x^2y)(-4xy^3)(x^2y^2)$

   (A) $8x^5y^6$
   (B) $-6x^5y^6$
   (C) $-5x^4y^6$
   (D) $8x^4y^6$
   (E) $-5x^5y^6$

5. Simplify: $(2a^2b^3)^5$

   (A) $10a^7b^8$
   (B) $32a^{10}b^{15}$
   (C) $10a^{10}b^{15}$
   (D) $10a^2b^3$
   (E) $32a^7b^8$

6. Simplify: $8a^2b-2ab(4b+a)$

   (A) $10a^2b-8ab^2$
   (B) $-2a^2b$
   (C) $8a^2b-8ab^2+a$
   (D) $-2a^5b^4$
   (E) $6a^2b-8ab^2$

7. Divide: $\dfrac{9x^2y-12xy^3}{3x^2y^2}$

   (A) $\dfrac{3}{y}-\dfrac{4y}{x}$
   (B) $3-12xy^2$
   (C) $-\dfrac{1}{x}$
   (D) $9xy-4y$
   (E) $-xy^2$

8. If $2x+5=3$, then $x+6$ equals

   (A) 7
   (B) 2
   (C) 5
   (D) $2\tfrac{1}{2}$
   (E) 10

9. Solve for x: $4(2x-1)=6x+2$

   (A) $1\tfrac{1}{2}$
   (B) -3
   (C) 5
   (D) 3
   (E) 7

10. Solve for x: $\dfrac{3}{4}x+\dfrac{x}{3}=13$

    (A) $30\tfrac{1}{3}$
    (B) 13
    (C) 39
    (D) 24
    (E) 12

11. Solve for x: $\dfrac{2x}{3} = \dfrac{3x-1}{4}$

   (A) $-\tfrac{1}{4}$
   (B) 3
   (C) 1
   (D) $\tfrac{1}{4}$
   (E) 0

12. Solve for x: $\sqrt{2x-1} = 5$

   (A) $5\tfrac{1}{2}$
   (B) 18
   (C) 13
   (D) 3
   (E) 12

13. Solve for x in terms of a, b, and c: $a = \dfrac{b}{1+cx}$

   (A) $\dfrac{b-a}{ac}$
   (B) $b-a-ac$
   (C) $\dfrac{b}{a+ac}$
   (D) $\dfrac{b}{a+c}$
   (E) $\dfrac{a-b}{ac}$

14. If $5x - 2y = 4$ and $x + y = 5$, then $x^2 + y$ equals

   (A) 5
   (B) 25
   (C) 3
   (D) 49
   (E) 7

15. If $\dfrac{x}{5} - \dfrac{x}{2} > 6$, then a possible value of x is

   (A) 20
   (B) -20
   (C) -21
   (D) -19
   (E) 19

16. Factor: $x^2 - 5x - 6$

   (A) $(x-3)(x+2)$
   (B) $(x-6)(x+1)$
   (C) $(x-6)(x-1)$
   (D) $(x+6)(x-1)$
   (E) $(x-5)(x+1)$

17. Factor: $9a^2 - 25b^4$

   (A) $(9a - 25b^2)^2$
   (B) $(9a - 25b^2)(9a + 25b^2)$
   (C) $(3a - 5b^2)^2$
   (D) $9(a - 5b^2)(a + 5b^2)$
   (E) $(3a + 5b^2)(3a - 5b^2)$

18. Factor: $2x^3 + 10x^2 - 48x$

   (A) $2x(x+8)(x-3)$
   (B) $2x(x-12)(x+2)$
   (C) $2(x-2)(x+8)(x-3)$
   (D) $2x(x-12)(x+4)$
   (E) $2x(x-8)(x+3)$

19. Solve for x: $x^2 + 2x = 35$

   (A) $8\tfrac{3}{4}$ or $-8\tfrac{3}{4}$
   (B) -5 or 7
   (C) $\sqrt{33}$ or $-\sqrt{33}$
   (D) -7 or 5
   (E) -5 or -7

20. Simplify: $\sqrt{50x^2y^3}$

   (A) $5x^2y^2\sqrt{2xy}$
   (B) $5xy\sqrt{2y}$
   (C) $5x^3y^3\sqrt{2}$
   (D) $25xy\sqrt{y}$
   (E) $10xy\sqrt{5y}$

21. Simplify: $\sqrt{2x} + \sqrt{32x} - \sqrt{18x}$

   (A) $4\sqrt{x}$
   (B) $8\sqrt{2x}$
   (C) $2\sqrt{2x}$
   (D) $4\sqrt{2x}$
   (E) $4x\sqrt{x}$

22. Simplify: $\dfrac{\sqrt{5xy} \cdot \sqrt{15x^3y}}{\sqrt{3}}$

   (A) $5x^2y$
   (B) $5x^4y^2$
   (C) $\dfrac{2xy\sqrt{5}}{\sqrt{3}}$
   (D) $\dfrac{5x^2y}{\sqrt{3}}$
   (E) $10xy$

23. Subtract: $\dfrac{5}{x-2} - \dfrac{2}{x+2}$

(A) $\dfrac{3}{4}$

(B) $\dfrac{3}{2x}$

(C) $\dfrac{3x+6}{x^2-4}$

(D) $\dfrac{3x}{x^2-4}$

(E) $\dfrac{3x+14}{x^2-4}$

24. Divide: $\dfrac{x^2-25}{x^2} \div \dfrac{3x+15}{x}$

(A) $\dfrac{x-5}{6x}$

(B) $\dfrac{x-5}{3x}$

(C) $\dfrac{x+5}{3}$

(D) $\dfrac{3x}{x-5}$

(E) $\dfrac{6x}{x-5}$

25. Simplify: $\dfrac{\dfrac{3}{x} + \dfrac{5}{y}}{2}$

(A) $\dfrac{6x+10y}{xy}$

(B) $\dfrac{4}{x+y}$

(C) $\dfrac{16}{x+y}$

(D) $\dfrac{3y+5x}{2xy}$

(E) $\dfrac{4}{xy}$

ALGEBRA PRACTICE TEST—ANSWER KEY

1. C	6. E	11. B	16. B	21. C
2. D	7. A	12. C	17. E	22. A
3. D	8. C	13. A	18. A	23. E
4. A	9. D	14. E	19. D	24. B
5. B	10. E	15. C	20. B	25. D

ALGEBRA PRACTICE TEST—ANSWERS AND SOLUTIONS

1.(C) $\quad 5^2 + 3(7-5)^3$
$= 5^2 + 3(2)^3$
$= 25 + 3(8)$
$= 25 + 24$
$= 49$

2.(D) $\quad \dfrac{\sqrt{b^2 - 4ax}}{a - x}$

$= \dfrac{\sqrt{(1)^2 - 4(2)(-3)}}{(2) - (-3)}$

$= \dfrac{\sqrt{1 + 24}}{2 + 3}$

$= \dfrac{\sqrt{25}}{5}$

$= \dfrac{5}{5}$

$= 1$

3.(D) $\quad (4x^2 + y^2) + (x^2 - 5y^2) - (2x^2 - y^2)$
$\qquad\qquad\qquad\qquad \downarrow \ \downarrow \ \downarrow$
$= (4x^2 + y^2) + (x^2 - 5y^2) + (-2x^2 + y^2)$

$\quad\quad\quad\quad 4x^2 + \ y^2$
$\quad\quad\quad\quad\ x^2 - 5y^2$
$\quad\quad\quad -2x^2 + \ y^2$
$\quad\quad\quad \overline{\ 3x^2 - 3y^2}$

4.(A) $\quad (-2x^2y)(-4xy^3)(x^2y^2)$
$= (-2)(-4)(x^2 \cdot x \cdot x^2)(y \cdot y^3 \cdot y^2)$
$= 8x^5y^6$

5.(B) $\quad (2a^2b^3)^5$
$= 2^5 a^{2 \cdot 5} b^{3 \cdot 5}$
$= 32 a^{10} b^{15}$

6.(E) $\quad 8a^2b - 2ab(4b + a)$
$= 8a^2b - 8ab^2 - 2a^2b$
$= 6a^2b - 8ab^2$

7.(A) $\quad \dfrac{9x^2y - 12xy^3}{3x^2y^2}$

$= \dfrac{9x^2y}{3x^2y^2} - \dfrac{12xy^3}{3x^2y^2}$

$= \dfrac{3}{y} - \dfrac{4y}{x}$

8.(C) $\quad 2x + 5 = \ \ 3$
$\quad\quad \dfrac{-5}{2x} \ \ \dfrac{-5}{= -2}$

$\quad\quad \dfrac{\cancel{2}x}{\cancel{2}} = \dfrac{-2}{2}$

$\quad\quad x = -1$

Thus, $x + 6 = -1 + 6$
$\quad\quad\quad\quad = 5$

9.(D) $\quad 4(2x - 1) = \ 6x + 2$
$\quad\quad\ 8x - 4 = \ 6x + 2$
$\quad\quad \dfrac{-6x}{2x - 4} \ \dfrac{-6x}{= \ +2}$
$\quad\quad \dfrac{+4}{2x} \ \dfrac{+4}{= \ +6}$

$\quad\quad \dfrac{\cancel{2}x}{\cancel{2}} = \dfrac{6}{2}$

$\quad\quad x = 3$

10.(E) $\quad \dfrac{3}{4}x + \dfrac{x}{3} = 13$ (multiply by 12)

$\cancel{12}^3 \cdot \dfrac{3}{\cancel{4}} x + \cancel{12}^4 \cdot \dfrac{x}{\cancel{3}} = 12 \cdot 13$

$9x + 4x = 156$

$\dfrac{\cancel{13}x}{\cancel{13}} = \dfrac{156}{13}$

$x = 12$

11.(B) $\quad \dfrac{2x}{3} = \dfrac{3x-1}{4}$ (cross-multiply)

$4(2x) = 3(3x-1)$

$\begin{aligned} 8x &= 9x - 3 \\ -9x & \quad -9x \\ \hline -x &= -3 \end{aligned}$

$x = 3$

12.(C) $\quad \sqrt{2x-1} = 5$ (square both sides)

$(\sqrt{2x-1})^2 = (5)^2$

$\begin{aligned} 2x - 1 &= 25 \\ +1 & \quad +1 \\ \hline 2x &= 26 \end{aligned}$

$\dfrac{\cancel{2}x}{\cancel{2}} = \dfrac{26}{2}$

$x = 13$

13.(A) $\quad a = \dfrac{b}{1+cx}$

$\dfrac{a}{1} = \dfrac{b}{1+cx}$ (cross-multiply)

$a(1+cx) = 1(b)$

$\begin{aligned} a + acx &= b \\ -a & \quad -a \\ \hline acx &= b - a \end{aligned}$

$\dfrac{\cancel{ac}x}{\cancel{ac}} = \dfrac{b-a}{ac}$

$x = \dfrac{b-a}{ac}$

14.(E) $\quad 5x - 2y = 4 \rightarrow 5x - 2y = 4$

$ 2(x+y=5) \rightarrow 2x + 2y = 10$

$ \overline{7x = 14}$

$\dfrac{\cancel{7}x}{\cancel{7}} = \dfrac{14}{7}$

$x = 2$ (substitute into $x + y = 5$)

$\begin{aligned} x + y &= 5 \\ 2 + y &= 5 \\ -2 & \quad -2 \\ \hline y &= 3 \end{aligned}$

Thus, $x^2 + y = 2^2 + 3$
$ = 4 + 3$
$ = 7$

15.(C) $\quad \dfrac{x}{5} - \dfrac{x}{2} > 6$ (multiply by 10)

$\cancel{10}^2 \cdot \dfrac{x}{\cancel{5}} - \cancel{10}^5 \cdot \dfrac{x}{\cancel{2}} > 10 \cdot 6$

$2x - 5x > 60$

$-3x > 60$
$\downarrow$

$\dfrac{-\cancel{3}x}{-\cancel{3}} < \dfrac{60}{-3}$ (reverse inequality)

$x < -20$

16.(B) $\quad x^2 - 5x - 6$

$= (x \quad)(x \quad)$

$= (x-6)(x+1) \qquad +x - 6x = -5x$

17.(E) $\quad 9a^2 - 25b^4$

$= \left(\sqrt{9a^2} + \sqrt{25b^4}\right)\left(\sqrt{9a^2} - \sqrt{25b^4}\right)$

$= (3a + 5b^2)(3a - b^2)$

18.(A) $\quad 2x^3 + 10x^2 - 48x$

$= 2x(x^2 + 5x - 24)$

$= 2x(x+8)(x-3) \qquad -3x + 8x = +5x$

19.(D) $\quad \begin{aligned} x^2 + 2x &= 35 \\ -35 & \quad -35 \\ \hline x^2 + 2x - 35 &= 0 \end{aligned}$

$(x+7)(x-5) = 0$

$\begin{aligned} x + 7 &= 0 \\ -7 & \quad -7 \\ \hline x &= -7 \end{aligned} \quad \bigg| \quad \begin{aligned} x - 5 &= 0 \\ +5 & \quad +5 \\ \hline x &= +5 \end{aligned}$

20.(B) $\quad \sqrt{50x^2y^3}$

$= \sqrt{25x^2y^2 \cdot 2y}$

$= \sqrt{25x^2y^2} \cdot \sqrt{2y}$

$= 5xy\sqrt{2y}$

21.(C) $\quad \sqrt{2x} + \sqrt{32x} - \sqrt{18x}$

$= \sqrt{2x} + \sqrt{16 \cdot 2x} - \sqrt{9 \cdot 2x}$

$= \sqrt{2x} + \sqrt{16} \cdot \sqrt{2x} - \sqrt{9} \cdot \sqrt{2x}$

$= \sqrt{2x} + 4\sqrt{2x} - 3\sqrt{2x}$

$= (1 + 4 - 3)\sqrt{2x}$

$= 2\sqrt{2x}$

22.(A) $\dfrac{\sqrt{5xy} \cdot \sqrt{15x^3y}}{\sqrt{3}}$

$= \sqrt{\dfrac{5xy \cdot 15x^3y}{3}}$

$= \sqrt{25x^4y^2}$

$= 5x^2y$

23.(E) $\dfrac{5}{x-2} - \dfrac{2}{x+2}$ (cross-multiply)

$= \dfrac{5(x+2) - 2(x-2)}{(x-2)(x+2)}$

$= \dfrac{5x+10-2x+4}{x^2-4}$

$= \dfrac{3x+14}{x^2-4}$

24.(B) $\dfrac{x^2-25}{x^2} \div \dfrac{3x+15}{x}$ (invert and multiply)

$= \dfrac{x^2-25}{x^2} \cdot \dfrac{x}{3x+15}$

$= \dfrac{\overset{1}{\cancel{(x+5)}}(x-5)}{\underset{x}{\cancel{x^2}}} \cdot \dfrac{\overset{1}{\cancel{x}}}{3\underset{1}{\cancel{(x+5)}}}$

$= \dfrac{x-5}{3x}$

25.(D) $\dfrac{\dfrac{3}{x} + \dfrac{5}{y}}{2}$ (multiply by xy)

$= \dfrac{\cancel{x}y \cdot \dfrac{3}{\cancel{x}} + x\cancel{y} \cdot \dfrac{5}{\cancel{y}}}{xy \cdot 2}$

$= \dfrac{3y+5x}{2xy}$

UNIT II. WORD PROBLEMS

In this unit, we will discuss a variety of word problems which have algebraic solutions. The problems discussed include

- arithmetic problems
- number, integer, and age problems
- average and mixture problems
- motion problems
- work problems
- set problems

In general, problem-solving involves four basic steps: (1) understanding the problem, (2) formulating a plan, (3) carrying out the plan, and (4) checking the solution.

Understanding the problem involves reading the problem carefully to determine what is given, what relationships and conditions must be satisfied, and, most important, what is unknown. In some problems, it may help to draw a diagram. Once it is determined what is unknown, suitable notation should be introduced. For example, an unknown number might be represented by the letter n or x, an unknown time by the letter t, an unknown distance by the letter d, and so on.

Formulating a plan involves recognizing some essential element in the problem that reminds you of another problem you have seen and solved before. Needless to say, the more problems you have solved of a similar nature, the easier it will be to make the appropriate connection. In some problems, it may help to focus on the unknown and to write down any formulas you know involving this type of quantity. In other problems, it may help to write equations which translate into symbolic form the relationships and conditions stated in words.

Carrying out the plan involves performing the specific operations and procedures necessary to obtain the solution. Whether multiplying fractions, or factoring quadratic equations, this step should be done carefully and accurately.

Checking the solution involves returning to the original problem with the solution obtained to determine whether or not it satisfies the given relationships and conditions.

9

9.1 FRACTION WORD PROBLEMS

In order to find a fractional part of a number, we multiply the fraction times the number. For example, to find $\frac{2}{3}$ of 18 we multiply $\frac{2}{3} \times 18$ and get the result 12.

Whole Part
↘ ↙
$\frac{2}{3}$ of 18 = 12

The number we take the fractional part *of* is called the **Whole,** and the result is called the **Part.**

$$\boxed{\text{Fraction} \times \text{Whole} = \text{Part}}$$

In most fraction word problems, we will be given values for two of the quantities in this equation, and will be asked to find the third. Note that, if we are given the Part and the Whole, and are asked to find the Fraction, we usually use the alternate form of the equation:

$$\boxed{\text{Fraction} = \frac{\text{Part}}{\text{Whole}}}$$

SAMPLE PROBLEMS

Problem 1: A team played M games and won N of them. What fraction of its games did it lose?

PROCEDURE	SOLUTION
(1) To find the number of games the team lost, subtract the number of games it won from the total number of games.	(1) Games Lost = M − N
(2) Form a fraction by placing the number of games lost over the total number of games played.	(2) Fraction = $\frac{\text{Part}}{\text{Whole}}$ $= \frac{M-N}{M}$ The team lost $\frac{M-N}{M}$ of its games. *Answer*

Problem 2: If a woman paints $\frac{2}{3}$ of her apartment one weekend, and her roommate paints $\frac{3}{4}$ of what remains the following weekend, what fraction still remains to be painted?

PROCEDURE	SOLUTION
(1) After the first weekend, $\frac{1}{3}$ of the apartment remains to be painted $(1-\frac{2}{3})$. To find the fraction painted by the roommate, multiply this remaining fraction by $\frac{3}{4}$.	(1) $\frac{3}{4}$ of $\frac{1}{3} = \frac{3}{4} \times \frac{1}{3}$ $= \frac{1}{4}$ painted by roommate
(2) Add the fractions painted by the woman and her roommate, and subtract the result from 1.	(2) $\frac{2}{3} = \frac{8}{12}$ painted by woman $+\frac{1}{4} = \frac{3}{12}$ painted by roommate $\frac{11}{12}$ painted together $1 - \frac{11}{12} = \frac{1}{12}$ remains

$\frac{1}{12}$ remains to be painted.   *Answer*

Problem 3: An automobile gasoline gauge reads $\frac{1}{8}$ full. After the gas tank is filled with 15 gallons, the gauge reads $\frac{7}{8}$ full. What is the capacity of the tank?

PROCEDURE	SOLUTION
(1) Represent the capacity of the tank by C.	(1) Let C = the capacity of the tank.
(2) To obtain the fraction of the tank filled, subtract the original gauge reading from the final gauge reading.	(2) $\frac{7}{8}$ $-\frac{1}{8}$ $\frac{6}{8} = \frac{3}{4}$ of the tank was filled
(3) Write an equation expressing the fact that $\frac{3}{4}$ of the capacity of the tank equals 15 gallons. Solve this equation for C.	(3) Fraction × Whole = Part $\frac{3}{4} \times C = 15$ $4 \cdot \frac{3}{4}C = 15 \cdot 4$ $3C = 60$ $\frac{3C}{3} = \frac{60}{3}$ $C = 20$ gallons

The capacity of the tank is 20 gallons.   *Answer*

Problem 4: A student spends $\frac{3}{4}$ of his money on tuition, and $\frac{1}{3}$ of what remains on books. If he has $50 left, how much money did he start with?

PROCEDURE	SOLUTION
(1) Represent the student's money at the start by S.	(1) Let S = the student's money at the start.
(2) After tuition, $\frac{1}{4}$ of the original amount of money remains $(1-\frac{3}{4})$. To find the fraction spent on books, multiply this remaining fraction by $\frac{1}{3}$.	(2) $\frac{1}{3}$ of $\frac{1}{4} = \frac{1}{3} \times \frac{1}{4}$ $= \frac{1}{12}$ spent on books
(3) Add the fractions spent on tuition and books, and subtract the result from 1.	(3) $\frac{3}{4} = \frac{9}{12}$ tuition $+ \frac{1}{12} = \frac{1}{12}$ books $\frac{10}{12} = \frac{5}{6}$ spent together $1 - \frac{5}{6} = \frac{1}{6}$ remains
(4) Write an equation expressing the fact that the fraction remaining times the original amount of money equals $50. Solve this equation for S.	(4) $\frac{1}{6}S = 50$ $6 \cdot \frac{1}{6}S = 50 \cdot 6$ $S = 300$ The student started with $300. *Answer*

PRACTICE PROBLEMS

Answers
(1) $\dfrac{M-2}{M+W-3}$
(2) $\dfrac{D}{20}$
(3) $\dfrac{3}{10}$
(4) 55 miles

(1) On a certain bus there are M men and W women. If 2 men and 1 woman get off the bus, what fraction of the remaining passengers are men?

(2) Five friends decide to rent a summer house at a total cost of D dollars. They agree to share the rent equally. Before the summer begins, 1 of the friends has to drop out. By how many dollars does the cost to each of the other friends increase?

(3) In company X, $\frac{1}{4}$ of the employees earn under $12,000 per year. If $\frac{3}{5}$ of the remaining employees earn between $12,000 and $18,000, what fraction of the employees earn more than $18,000?

(4) After traveling 220 miles, Hilda completed $\frac{4}{5}$ of her trip. How far does she still have to travel?

(5) A family budgets $\frac{1}{3}$ of its monthly income for rent, $\frac{1}{4}$ for food, and $\frac{1}{6}$ for clothing. If, after paying these items, the family has $420 left, what is the monthly income?

> Answers
> (5) $1680

9.2 PROPORTION PROBLEMS

In proportion problems, we are told the rate at which two quantities vary, and are asked to find the value of one of them, given the value of the other.

For example, suppose we are told that a certain item costs $5 per dozen, and we are asked to find out how much 84 of them would cost. By expressing the cost rate as a ratio, we can write the following proportion: $5 is to 12 items (per dozen) as $x is to 84 items, where x represents the unknown cost. That is,

$$\frac{\text{Cost}}{\text{\# Items}} \rightarrow \frac{\$5}{12} = \frac{\$x}{84}$$

Notice that the order of the quantities on both sides of the proportion is the same—cost to number of items.

In order to find the unknown cost, x, we cross-multiply the numerators and denominators, set the products equal, and divide.

Thus, we get

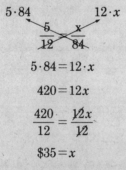

$$5 \cdot 84 = 12 \cdot x$$
$$420 = 12x$$
$$\frac{420}{12} = \frac{\cancel{12}x}{\cancel{12}}$$
$$\$35 = x$$

In other words, at the rate of $5 per dozen, 84 items would cost $35.

SAMPLE PROBLEMS

Problem 5: A recipe calls for M cups of water for every N cups of rice. How many cups of water are needed to make Q cups of rice?

PROCEDURE	SOLUTION
(1) Represent the unknown number of cups of water by x.	(1) Let $x =$ the number of cups of water needed to make Q cups of rice.

Arithmetic Problems / 93

(2) Write a proportion in which the number of cups of water is compared to the number of cups of rice. Remember to keep the order of the two quantities the same on both sides of the proportion. Solve this proportion for x by cross-multiplying and dividing.

(2) $\dfrac{\text{Water}}{\text{Rice}} \begin{array}{c}\rightarrow\\ \rightarrow\end{array} \dfrac{M}{N} = \dfrac{x}{Q}$

$Q \cdot M = N \cdot x$

$\dfrac{QM}{N} = \dfrac{\cancel{N}x}{\cancel{N}}$

$\dfrac{QM}{N} = x$

The number of cups of water needed to make Q cups of rice is $\dfrac{QM}{N}$. *Answer*

Problem 6: In a certain class, 4 out of every 5 students passed an exam. If 20 students passed the exam, how many failed?

PROCEDURE

(1) Represent the total number of students in the class by T.

(2) Write a proportion in which the number of people that passed the exam are compared to the total number of students. Solve this proportion for T by cross-multiplying and dividing.

(3) To find the number of students that failed the exam, subtract the number of students that passed the exam, 20, from this total.

SOLUTION

(1) Let T = the total number of students in the class.

(2) $\dfrac{\text{Passed}}{\text{Total}} \begin{array}{c}\rightarrow\\ \rightarrow\end{array} \dfrac{4}{5} = \dfrac{20}{T}$

$4 \cdot T = 5 \cdot 20$

$\dfrac{\cancel{4}T}{\cancel{4}} = \dfrac{100}{4}$

$T = 25$

(3) Failed = Total − Passed

Failed = 25 − 20 = 5

The number of students that failed the exam is 5. *Answer*

PRACTICE PROBLEMS

Answers

(1) $\dfrac{45N}{M}$
(2) 90 miles
(3) $\dfrac{25DR}{3N}$

(1) A team wins M out of every N games played. If the team wins 45 games, how many games does it play?

(2) The scale on a map is $\frac{3}{4}$ inch = 15 miles. If the distance between two cities is represented by $4\frac{1}{2}$ inches on the map, what is the actual distance between the cities?

(3) If N dozen apples cost D dollars, what is the cost, in *cents*, of R apples at the same rate?

(4) Mary can read P pages in M minutes. At the same rate, how many *hours* will it take her to read a book of 300 pages?

(5) In a recent survey, R out of S people said they walked to work. If 50 people said they walked to work, how many people said they did not walk to work?

> **Answers**
>
> (4) $\dfrac{5M}{P}$
>
> (5) $\dfrac{50(S-R)}{R}$

9.3 PERCENT WORD PROBLEMS—GENERAL

Most percent problems involve three quantities—a *Percent*, a *Whole*, and a *Part*. These quantities are related by the following product and proportion:

Percent Product	Percent Proportion
Percent × Whole = Part	$\dfrac{\text{Part}}{\text{Whole}} = \dfrac{\text{Percent}}{100\%}$

In percent word problems, we will be given values for two of these quantities, and will be asked to find the value of the third. Either the product or the proportion can be used. Remember that if the product is used, the Percent must be changed to either a fraction or a decimal.

SAMPLE PROBLEMS

Problem 7: Of its net monthly income, a family budgets 25% for rent, and 20% for food. If its net monthly income is $1200, how much money is budgeted for things other than rent and food?

PROCEDURE	SOLUTION
(1) Represent the amount of money budgeted for other things by N.	(1) Let N = the amount of money budgeted for things other than rent and food.

(2) To find the percent budgeted for other things, add the two given percents and subtract the result from 100%.	(2) 25% rent
	  +20% food
	    45%
	
	100% − 45% = 55% budgeted for other things

(3) Using either the product or the proportion, solve for N. With the product, remember to change the percent to a decimal (D←P). With the proportion, cross-multiply and divide.

(3)   Product

$$\% \times \text{Whole} = \text{Part}$$

$$.55 \times 1200 = N$$

$$660 = N$$

Proportion

$$\frac{\text{Part}}{\text{Whole}} = \frac{\%}{100\%}$$

$$\frac{N}{1200} = \frac{55}{100}$$

$$100 \cdot N = 55 \cdot 1200$$

$$100N = 66000$$

$$\frac{\cancel{100}N}{\cancel{100}} = \frac{66000}{100}$$

$$N = 660$$

The amount budgeted for things other than rent and food is $660.   *Answer*

Problem 8: On a certain test, 18 students passed and 7 students failed. What percent passed the test?

PROCEDURE

(1) Represent the percent that passed by P.

(2) Using either the product or the proportion, solve for P. In both methods, the Whole is 25 (18+7). With the product, remember to move the decimal point in the result two places to the right (D→P).

SOLUTION

(1) Let P = the percent that passed.

(2)   Product

$$\% \times \text{Whole} = \text{Part}$$

$$P \times 25 = 18$$

$$\frac{\cancel{25}P}{\cancel{25}} = \frac{18}{25}$$

$$P = .72$$

$$P = 72\%$$

Proportion

$$\frac{\text{Part}}{\text{Whole}} = \frac{\%}{100\%}$$

$$\frac{18}{25} = \frac{P}{100}$$

$$25 \cdot P = 18 \cdot 100$$

$$\frac{\cancel{25}P}{\cancel{25}} = \frac{1800}{25}$$

$$P = 72\%$$

72% passed the test.   *Answer*

Problem 9: A woman gave $115 as a deposit on a stereo system. If this represents 20% of the total cost, how much does she still owe?

PROCEDURE

(1) Represent the total cost of the stereo system by C.

(2) Using either the product or the proportion, solve for C. With the product, remember to change the percent to a decimal (D←P).

SOLUTION

(1) Let C = the total cost of the stereo system.

(2)   Product

$$\% \times \text{Whole} = \text{Part}$$

$$.20 \times C = 115$$

$$\frac{.\cancel{20}C}{.\cancel{20}} = \frac{115}{.20}$$

$$C = 575$$

Proportion

$$\frac{\text{Part}}{\text{Whole}} = \frac{\%}{100\%}$$

$$\frac{115}{C} = \frac{20}{100}$$

$$20 \cdot C = 115 \cdot 100$$

$$\frac{\cancel{20}C}{\cancel{20}} = \frac{11500}{20}$$

$$C = 575$$

(3) To find the amount still owed, subtract the amount deposited from the total cost.

(3)
$575 total cost
−$115 deposit
─────
$460

The woman still owes $460.   *Answer*

PRACTICE PROBLEMS

(1) Rachel invests $3000 at 8% simple annual interest. How much interest is earned after 3 months?

(2) In an election, only R, S, and T received votes. If S received twice as many votes as R, and T received three times as many votes as S, what percent of the total vote was received by T?

(3) In a class of 40 students, 80% are men. Of the 24 students that pass the midterm, 75% are men. What percent of the women in the class pass the midterm?

(4) After having dinner at a restaurant, Anna leaves $6.30 as a tip for the waiter. If this represents 15% of the total bill, how much was the bill?

(5) A manufacturer finds that, on the average, .2% of his items must be rejected. If, in a certain month, 5 items are rejected, how many items pass the inspection?

(6) A car dealer advertises two different payment plans for a new car. If the buyer pays cash, the car costs $5700. If the buyer pays on the installment plan, he pays 20% of the cash cost as a down payment, and then $200 a month for 24 months. How much more money must a buyer pay on the installment plan than on the cash plan?

(7) R percent of what number is Q?

(8) In a certain high school graduating class, there are 200 boys and 300 girls. If 60% of the boys and 70% of the girls go to college, what percent of the entire graduating class go to college?

(9) After traveling 480 miles, Steve completed 80% of his trip. How much further must he still travel?

(10) If 30% of 50% of $70x$ equals 21, what is the value of x?

Answers	
(1)	$60
(2)	$66\frac{2}{3}\%$
(3)	75%
(4)	$42
(5)	2495
(6)	$240
(7)	$\frac{100Q}{R}$
(8)	66%
(9)	120 miles
(10)	2

9.4 PERCENT WORD PROBLEMS—PERCENT OF INCREASE OR DECREASE

Sometimes, we are asked to find the percent of *change* (increase or decrease) from one number to another number. In these problems, we use the special forms of the percent product and proportion shown below. Notice that in both forms, the *Part* is represented by the *Amount of Change*, and the *Whole* is represented by the *Original Value*.

Percent of Change Product

% of Change $\times$ Original Value = Amount of Change

Percent of Change Proportion

$$\frac{\text{Amount of Change}}{\text{Original Value}} = \frac{\text{\% of Change}}{100\%}$$

SAMPLE PROBLEM

Problem 10: A class increased in size from 20 students to 25 students. What was the percent of increase?

PROCEDURE	SOLUTION
(1) Represent the percent of increase by P.	(1) Let P = the percent of increase
(2) To find the amount of increase, subtract the original class size from the new class size.	(2) Amount of Increase = 25 − 20 = 5
(3) Using either the product or proportion, solve for P.	(3) *Product* $\qquad$ *Proportion*

$$\text{\% Inc.} \times \text{Orig.} = \text{Inc.} \qquad \frac{\text{Inc.}}{\text{Orig.}} = \frac{\text{\% Inc.}}{100\%}$$

$$P \times 20 = 5 \qquad\qquad \frac{5}{20} = \frac{P}{100}$$

$$\frac{20P}{20} = \frac{5}{20} \qquad\quad 20 \cdot P = 5 \cdot 100$$

$$P = .25 \qquad\qquad \frac{20P}{20} = \frac{500}{20}$$

$$P = 25\% \qquad\qquad\quad P = 25\%$$

The class increased 25% in size. $\qquad$ *Answer*

PRACTICE PROBLEMS (A)

(1) Before starting a diet, Norm weighed 180 pounds. After the diet, he weighed 153 pounds. What was his percent of weight loss?

(2) The number of employees at company X increases from 50 to 60. What is the percent of increase?

(3) Patty sold her camera for $300, thus making a $60 profit. What was her percent of profit?

(4) A piano, which normally sells for $800, is marked down to $700. What is the percent of markdown?

(5) Marta buys a house for C dollars, and later sells it for R dollars, where R is greater than C. What is her percent of profit?

Answers
(1) 15%
(2) 20%
(3) 25%
(4) $12\tfrac{1}{2}\%$
(5) $\dfrac{100(R-C)}{C}$

When we are given the original value of a quantity, and the percent of increase or decrease, we can find the new value by two different methods.

Method 1. We use the percent product to find the amount of increase or decrease, and then add it to (in the case of an increase), or subtract it from (in the case of a decrease), the original value.

> *% Increase*: New Value = Original Value + Amount of Increase
>
> *% Decrease*: New Value = Original Value − Amount of Decrease

Method 2. We add the percent of increase to 100%, or subtract the percent of decrease from 100%, and then multiply the resulting percent by the original value.

> *% Increase*: New Value = (100% + % Increase) × Original Value
>
> *% Decrease*: New Value = (100% − % Decrease) × Original Value

For example, if the value of an item is increased by 25%, then its new value is 100% + 25%, or 125%, of its original value. Similarly, if the value of an item is reduced by 25%, then its new value is 100% − 25%, or 75%, of its original value.

The second method is particularly useful when applied to business problems involving a profit or a discount:

> *% Profit*:   Selling Price = (100% + % Profit) × Cost
>
> *% Discount*: Selling Price = (100% − % Discount) × List Price

SAMPLE PROBLEMS

Problem 11: A certain stock decreased in value by 15%. If it was originally selling at $80 a share, what is its value after the decrease?

PROCEDURE

Method 1

(1) To find the amount of decrease, multiply the percent of decrease by the original value.

(2) Subtract the amount of decrease from the original value.

SOLUTION

(1) Amount of Decrease = % Dec. × Orig. Value
= .15 × $80
= $12

(2) New Value = Orig. Value − Amount of Decrease
= $80 − $12
= $68   *Answer*

Method 2

(1) Subtract the percent of decrease from 100%, and multiply the resulting percent by the original value.

(1) New Value = (100% − 15%) × $80
= (85%) × $80
= .85 × $80
= $68

The new value of the stock is $68.   *Answer*

Problem 12: A music store sells a guitar for $480, thus making a profit of 20% of their cost. What did the guitar cost the store?

PROCEDURE

(1) Represent the cost by C.

(2) Write an equation expressing the fact that the selling price is equal to 120% of the cost. Solve this equation for C.

SOLUTION

(1) Let C = the cost of the guitar for the store.

(2) S. Price = (100% + 20%) × Cost
$480 = (120%) × C
$480 = 1.20C
$$\frac{\$480}{1.20} = \frac{1.20C}{1.20}$$
$400 = C

The guitar cost the store $400.   *Answer*

Problem 13: During a sale, a camera is marked 25% off the list price. If the camera is on sale for $330, what is the list price?

PROCEDURE	SOLUTION
(1) Represent the list price by L.	(1) Let L = the list price of the camera.
(2) Write an equation expressing the fact that the sales price is 75% of the list price. Solve this equation for L.	(2) S. Price = (100% − 25%) × List Price $$\$330 = (75\%) \times L$$ $$\$330 = .75L$$ $$\frac{\$330}{.75} = \frac{.75L}{.75}$$ $$\$440 = L$$ The list price of the camera is $440. *Answer*

PRACTICE PROBLEMS (B)

(1) If a $300 typewriter loses 10% of its value each year, what is its value after three years?

(2) Stu sold his cassette deck for $600, thus making a profit of 25%. What did the cassette deck originally cost Stu?

(3) The population of a certain city is 44,800, an increase of 12% from the previous year. What was the population in the previous year?

(4) An airline offers a special night flight at a 30% discount off the day flight fare. If the night flight costs $280, what is the day flight fare?

(5) A store buys an item for $40. At what price should the store list the price of the item in order to give its customers a 20% discount off the list price, and still make a 10% profit on the cost?

Answers
(1) $218.70
(2) $480
(3) 40,000
(4) $400
(5) $55

10

10.1 NUMBER PROBLEMS INVOLVING ONE UNKNOWN

In number problems involving one unknown quantity, we are asked to find a number which satisfies a given condition. This condition can usually be expressed as an algebraic equation by translating the words and phrases in the condition into symbolic form.

Following are the most common words and phrases, along with their translations into symbolic form:

WORD OR PHRASE	SYMBOLIC FORM
the *sum* of x and 5, x *plus* 5, x *increased by* 5, 5 *more than* x	$x+5$
the *difference* of x and 2, x *minus* 2, x *decreased by* 2, x *less* 2, 2 *less than* x (note the reversal)	$x-2$
the *product* of 7 and x, 7 *times* x	$7x$
$\frac{3}{4}$ *of* x (a fraction *of* x), 75% *of* x (a percent *of* x)	$\frac{3}{4}x$, $.75x$
the *quotient* of x and 3, x *divided by* 3, 3 *divided into* x (note the reversal)	$\frac{x}{3}$
x *equals* 9, x *is the same as* 9, x *is* 9	$x=9$

SAMPLE PROBLEMS

Problem 14: If 4 times a number is increased by 7, the result is 1 less than 6 times the same number. What is the number?

PROCEDURE	SOLUTION	
(1) Represent the unknown number by n.	(1) Let $n=$ the number.	
(2) Translate the given condition into symbolic form. Note that "1 *less than* $6n$" is written $6n-1$.	(2) "If 4 times a number is increased by 7,	$4n+7$
	the result is	$=$
	1 less than 6 times the same number."	$6n-1$

(3) Solve the resulting equation for n.

(3)
$$\begin{aligned} 4n+7 &= 6n-1 \\ -4n & -4n \\ \hline 7 &= 2n-1 \\ +1 & +1 \\ \hline 8 &= 2n \\ \frac{8}{2} &= \frac{2n}{2} \\ 4 &= n \end{aligned}$$

The number is 4.   *Answer*

Problem 15: If 9 more than a number equals 3 times the sum of the number and 1, find the number.

PROCEDURE	SOLUTION

(1) Represent the unknown number by n.
(2) Translate the given condition into symbolic form. Note that "3 times the sum" must be written with parentheses.

(1) Let n = the number.
(2)

"If 9 more than a number	$n+9$
equals	$=$
3 times the sum of the number and 1"	$3(n+1)$

(3) Solve the resulting equation for n.

(3)
$$\begin{aligned} n+9 &= 3(n+1) \\ n+9 &= 3n+3 \\ -n & -n \\ \hline 9 &= 2n+3 \\ -3 & -3 \\ \hline 6 &= 2n \\ \frac{6}{2} &= \frac{2n}{2} \\ 3 &= n \end{aligned}$$

The number is 3.   *Answer*

PRACTICE PROBLEMS

(1) If 6 times a number is decreased by 4, the result is the same as when 3 times the number is increased by 2. What is the number?

(2) The sum of a number and twice the same number is equal to 8 less than 5 times the number. What is the number?

(3) Five less than twice a number is equal to one-half the difference of the number and 1. What is the number?

(4) If the quotient of a number and 3 is decreased by 2, the result is equal to 1 more than the quotient of the number and 4. What is the number?

(5) If $\frac{3}{4}$ of a number is decreased by 5, the result is equal to $\frac{1}{3}$ of the same number. What is the number?

Answers
(1) 2
(2) 4
(3) 3
(4) 36
(5) 12

10.2 NUMBER PROBLEMS INVOLVING MORE THAN ONE UNKNOWN

In number problems involving more than one unknown, we are asked to find several numbers which satisfy a given condition. By using information given in the problem, we can usually represent all of the unknowns in terms of one variable, instead of using a different variable for each.

SAMPLE PROBLEMS

Problem 16: The larger of two numbers is 3 more than twice the smaller. If their sum is 18, what are the numbers?

PROCEDURE	SOLUTION
(1) Represent the smaller number by s, and the larger number by $2s+3$ ("3 more than twice the smaller").	(1) Let s = the smaller number, and $2s+3$ = the larger number.
(2) Translate the given condition into symbolic form.	(2) "If their sum is 18" → $s+(2s+3) = 18$
(3) Solve the equation for s, and substitute the result into the expression for the larger number.	(3) $s+(2s+3) = 18$ $3s+3 = 18$ $-3 -3$ $3s = 15$ $\dfrac{3s}{3} = \dfrac{15}{3}$ $s = 5$ Thus, $2s+3 = 2(5)+3$ $ = 13$ The smaller number is 5, and the larger number is 13. *Answer*

Problem 17: A class has 38 students. If 3 less than the number of men equals 5 more than twice the number of women, how many women are in the class?

PROCEDURE	SOLUTION
(1) Represent the number of men by M. Since there are a total of 38 people in the class, this leaves $38-M$ women.	(1) Let M = the number of men, and $38-M$ = the number of women.
(2) Translate the given condition into symbolic form.	(2) "If 3 less than the number of men" → $M-3$ "equals" → $=$ "5 more than twice the number of women" → $2(38-M)+5$

(3) Solve the equation for M, and substitute the result into the expression for the number of women.

(3)
$$M - 3 = 2(38 - M) + 5$$
$$M - 3 = 76 - 2M + 5$$
$$M - 3 = 81 - 2M$$
$$\underline{+2M \qquad\qquad +2M}$$
$$3M - 3 = 81$$
$$\underline{+3 \quad +3}$$
$$3M = 84$$
$$\frac{3M}{3} = \frac{84}{3}$$
$$M = 28$$

Thus, $38 - M = 10$

The number of women is 10.   *Answer*

PRACTICE PROBLEMS

(1) The smaller of two numbers is 12 less than three times the larger. If their difference is 2, what is the larger number?

(2) A carpenter cuts a board, 36 inches long, into two pieces, such that twice the longer piece is 2 inches more than three times the shorter piece. How many inches is the longer piece?

(3) An inheritance of $18,000 is divided among three children in the ratio of 2:3:4. How much is the largest share?

(4) Nancy invested part of $3000 at 12% simple annual interest, and the rest at 9% simple annual interest. If the total interest earned after one year was $330, how much did she invest at the 12% rate?

(5) At the 5 P.M. showing of a recent hit movie, 300 tickets were sold for a total of $1000. If people with student discount tickets paid one-half the regular price of $5.00, how many student discount tickets were sold?

Answers
(1) 5
(2) 22
(3) $8000
(4) $2000
(5) 200

10.3 CONSECUTIVE INTEGER PROBLEMS

Consecutive integers are integers (positive whole numbers, negative whole numbers, and 0) which differ by 1. Algebraically, we can represent them by N, N+1, N+2, etc., where N is any integer.

Consecutive even integers and **consecutive odd integers** are integers which differ by 2. Algebraically, we can represent them both by N, N+2, N+4, etc. Note that for consecutive *even* integers, N must be *even*, and for consecutive *odd* integers, N must be *odd*.

Consecutive Integers
$\ldots, -2, -1, 0, 1, 2, \ldots$

Consecutive Even
$\ldots, -4, -2, 0, 2, 4, \ldots$

Consecutive Odd
$\ldots, -3, -1, 1, 3, 5, \ldots$

SAMPLE PROBLEMS

Problem 18: Find three consecutive integers whose sum is 21.

PROCEDURE	SOLUTION

(1) Represent the three consecutive integers algebraically.

(1) Let N = the first integer,
$N+1$ = the second integer,
and $N+2$ = the third integer.

(2) Translate the given condition into symbolic form.

(2)

"Their sum	$N+(N+1)+(N+2)$
is	=
21"	21

(3) Solve the equation for N, and substitute the result into the expressions for the other two integers.

(3)
$$N+(N+1)+(N+2) = 21$$
$$3N+3 = 21$$
$$\frac{-3 \quad -3}{3N \quad = 18}$$
$$\frac{\cancel{3}N}{\cancel{3}} = \frac{18}{3}$$
$$N = 6$$

Thus, $N+1 = 7$, and $N+2 = 8$.

The integers are 6, 7, and 8.   *Answer*

Problem 19: Find three consecutive *even* integers such that 5 times the first, minus twice the second, equals the third.

PROCEDURE	SOLUTION

(1) Represent the three consecutive *even* integers algebraically.

(1) Let N = the first even integer,
$N+2$ = the second even integer,
and $N+4$ = the third even integer.

(2) Translate the given condition into symbolic form.

(2)

"5 times the first, minus twice the second	$5N-2(N+2)$
equals	=
the third"	$N+4$

(3) Solve the equation for N, and substitute the result into the expressions for the other two integers.

(3) $5N - 2(N+2) = N+4$
$5N - 2N - 4 = N + 4$

$$\begin{aligned} 3N - 4 &= N + 4 \\ -N & \quad -N \\ \hline 2N - 4 &= \quad 4 \\ +4 & \quad +4 \\ \hline 2N &= \quad 8 \end{aligned}$$

$$\frac{2N}{2} = \frac{8}{2}$$

$$N = 4$$

Thus, $N + 2 = 6$, and $N + 4 = 8$.

The even integers are 4, 6, and 8.   *Answer*

PRACTICE PROBLEMS

(1) Find the smallest of three consecutive even integers whose sum is 48.

(2) If the sum of three consecutive integers is 13 more than the smallest, what is their sum?

(3) Find the smallest of three consecutive odd integers such that the sum of the first and the second is 27 less than 3 times the third.

(4) Find the smallest of three consecutive integers such that 8 more than the product of the first and the second is equal to the product of the second and the third.

(5) If the product of two consecutive *positive* odd integers is 63, what is their sum?

Answers
(1) 14
(2) 18
(3) 17
(4) 3
(5) 16

10.4 AGE PROBLEMS

In age problems, we are asked to find the present ages of certain people, given information comparing their ages in the present, as well as their ages at a specified time in the future or the past.

Remember that to represent an age N years in the future, we add N to the present age, and to represent an age N years in the past, we subtract N from the present age.

> Age N years in the future = Present age + N
> Age N years in the past = Present age − N

SAMPLE PROBLEMS

Problem 20: A father is now 3 times as old as his daughter. In 12 years he will be twice as old as his daughter will be then. What are their present ages?

PROCEDURE	SOLUTION
(1) Represent the daughter's present age by D, and the father's present age by 3D ("3 times as old").	(1) Let D = the daughter's present age, and $3D$ = the father's present age.
(2) To represent their ages 12 years in the future, add 12 to their present ages.	(2) $D+12$ = the daughter's age in 12 years. $3D+12$ = the father's age in 12 years.
(3) Translate the given condition into symbolic form.	(3) "In 12 years he will be twice as old as his daughter will be then" → $3D+12 = 2(D+12)$
(4) Solve the equation for D, and substitute the result into the expression for the father's present age.	(4) $3D+12 = 2(D+12)$ $$\begin{aligned}3D+12 &= 2D+24\\ -2D &\ -2D\\ \hline D+12 &= +24\\ -12 &\ -12\\ \hline D &= 12\end{aligned}$$ $D = 12$ Thus, $3D = 36$

The daughter's present age is 12, and the father's present age is 36.   *Answer*

Problem 21: A man is now 42 years old and his friend is 33 years old. How many years ago was the man twice as old as his friend was then.

PROCEDURE	SOLUTION
(1) Represent the "number of years ago" by N.	(1) Let N = the number of years ago.
(2) To represent the ages of the man and his friend N years ago, subtract N from their present ages.	(2) $42-N$ = the age of the man N years ago. $33-N$ = the age of the friend N years ago.
(3) Translate the given condition into symbolic form.	(3) "the man was twice as old as his friend was then" → $42-N = 2(33-N)$

(4) Solve the equation for N.

(4)
$$42 - N = 2(33 - N)$$

$$
\begin{aligned}
42 - N &= 66 - 2N \\
+2N &= +2N \\
\hline
42 + N &= 66 \\
-42 & -42 \\
\hline
N &= 24
\end{aligned}
$$

The number of years ago is 24.   **Answer**

PRACTICE PROBLEMS

(1) Sara is now 3 times older than Kristin. Four years ago, Sara was 5 times as old as Kristin was then. How old is Sara now?

(2) Eileen is now 24 years older than her daughter. In 8 years, Eileen will be twice as old as her daughter will be then. How old is her daughter now?

(3) James is now 3 years older than Lynette. If 7 years from now the sum of their ages will be 79, how old is Lynette now?

(4) Anne is 25 years old, and Francis is 21 years old. How many years ago was Anne 3 times as old as Francis was then?

(5) The sum of Hank's age and Eloise's age is 60. If Hank's age 8 years from now will be 3 times Eloise's age 4 years ago, how old is Hank now?

Answers	
(1)	24
(2)	16
(3)	31
(4)	19
(5)	40

11

11.1 AVERAGE PROBLEMS—SIMPLE AVERAGE

The **average**, or **mean**, of a set of numbers is equal to the sum of the numbers divided by the number of terms in the set. In other words, the average of the set of N numbers, $a_1, a_2, a_3, \ldots, a_N$, is given by the equation

$$\text{Average} = \frac{a_1 + a_2 + a_3 + \cdots + a_N}{N}$$

SAMPLE PROBLEMS

Problem 22: If a woman bowls scores of 187, 193, and 211 in her first three games, what must she score in her final game in order to average 200 overall?

PROCEDURE	SOLUTION
(1) Represent the score of her final game by S.	(1) Let S = the score of the final game.
(2) Using the definition of average, divide the sum of the scores by the number of scores. Solve the resulting equation for S.	(2) $\text{Average} = \frac{\text{Sum}}{N}$ $200 = \frac{187 + 193 + 211 + S}{4}$ $200 = \frac{591 + S}{4}$ $4(200) = 4\left(\frac{591 + S}{4}\right)$ $800 = 591 + S$ $\underline{-591 \quad -591}$ $209 = S$ $209 = S$ The final score must be 209. *Answer*

Problem 23: During a one week period in January there were 4 days of snow. If the average daily snowfall for those 4 days was .35 inches, what was the average daily snowfall for the entire week?

PROCEDURE	SOLUTION
(1) Represent the total snowfall by T.	(1) Let T = the total number of inches of snowfall during the 4 days.
(2) Using the definition of average, divide the total snowfall, T, by the number of days it snowed, 4. Solve the resulting equation for T.	(2) $\text{Average} = \frac{\text{Sum}}{N}$ $.35 = \frac{T}{4}$ $4(.35) = 4\left(\frac{T}{4}\right)$ $1.40 \text{ in.} = T$
(3) To find the average for the entire week, divide the total snowfall by 7 (7 days in a week).	(3) $\text{Average} = \frac{\text{Sum}}{N}$ $= \frac{1.40}{7}$ $= .2$ The average daily snowfall for the week was .2 inches. *Answer*

PRACTICE PROBLEMS

(1) Barbara has marks of 92%, 86%, and 89% on her first three tests. What mark must she get on her next test in order to have a 90% average overall?

(2) The average weight of Don, Pat, and Jennifer is 135 lbs. How much does Adrian weigh if the average weight of all four people is 120 lbs?

(3) In the first game of a bowling tournament, a three-person team averaged 190. If, in the second game, one person's score increased by 7, another person's score increased by 15, and the third person's score decreased by 4, what was the teams's average score in the second game?

(4) The average of two numbers is A. If one of the numbers is P, what is the other number?

(5) The average of four numbers is what percent of the sum of the four numbers?

Answers
(1) 93%
(2) 75 lbs.
(3) 196
(4) 2A − P
(5) 25%

11.2 AVERAGE PROBLEMS—WEIGHTED AVERAGE

To combine the averages of different sets of numbers, the average of each set must be "weighted" (multiplied) by the number of terms in that set. For example, if a class of 12 students has an average of 80% on a test, and another class of 18 students has an average of 90% on the same test, then the average of the two classes combined is the sum of all the scores (12 with an average of 80% and 18 with an average of 90%) divided by the total number of scores (12+18=30).

$$\text{Combined Avg.} = \frac{12(80\%) + 18(90\%)}{12 + 18}$$

$$= \frac{2580\%}{30}$$

$$= 86\%$$

In this example, the numbers 12 and 18 are referred to as the **weights** for the averages 80% and 90%, respectively. Notice that, since there are more scores averaging 90% than 80%, the combined average is higher than 85%, the simple average of 80% and 90%.

In general,

The **combined (weighted) average** of a group of N_1 numbers, having an average of A_1, and another group of N_2 numbers, having an average of A_2, is

$$\text{Combined Avg.} = \frac{N_1 \cdot A_1 + N_2 \cdot A_2}{N_1 + N_2}$$

SAMPLE PROBLEM

Problem 24: According to the table below, what is the average number of credits taken per semester by the freshmen and sophomores combined, rounded off to the nearest tenth?

	Number of Students	Avg. Number of Credits/Semester
Freshmen	300	12.5
Sophomores	200	14.8

PROCEDURE	SOLUTION
(1) Multiply each average by the corresponding number of students, and add the results together. Divide the sum by the total number of students. Round off the result to the nearest tenth.	(1) Comb. Avg. $= \dfrac{300(12.5)+200(14.8)}{300+200}$ $= \dfrac{6710}{500}$ $= 13.42$ ≈ 13.4 The average number of credits taken per semester by the freshmen and sophomores combined is 13.4, rounded off to the nearest tenth. *Answer*

PRACTICE PROBLEMS

(1) In a certain school, 2 teachers earn $16,000 per year, 3 teachers earn $17,000 per year, and 5 teachers earn $18,000 per year. What is the average annual salary of these teachers?

(2) On the midterm exam, the average grade of a class was 86. If 20% of the class averaged 95, and 30% of the class averaged 90, what was the average grade of the rest of the class?

(3) The average temperature for the first 4 days of a week was 50°. If the average temperature for the next 3 days was 43°, what was the average temperature for the entire week?

(4) The average age of 10 friends is 32 years. If 3 of them average 28 years, and 5 others average 34 years, what is the average age of the remaining 2 people?

(5) The average of 8 scores is Q. If the average of 3 of the scores is Z, what is the average of the other 5 scores?

Answers
(1) $17,300
(2) 80
(3) 47°
(4) 33 years
(5) $\dfrac{8Q-3Z}{5}$

11.3 MIXTURE PROBLEMS

In mixture problems, we are given the unit value of several different items, and are asked to determine the number of units that should be used of each to obtain a mixture having a specified value.

The *value of each item* in a mixture is equal to the number of units of that item multiplied by the value per unit (the unit value). For example, the value of 3 pounds of coffee, worth $2.25 per pound, is 3($2.25), or $6.75. In general,

> The Value of Each Item = (Number of Units) · (Unit Value)

The *value of the mixture* is equal to the *sum* of the values of the items within it. For example, the value of a mixture of 5 pounds of nuts, worth $2.00 per pound, and 4 pounds of candy, worth $3.00 per pound, is 5($2.00)+4($3.00), or $22.00.

In general,

> If N_1 units of an item having a unit value of U_1 are mixed with N_2 units of an item having a unit value of U_2, then
> The Value of the Mixture = $N_1 \cdot U_1 + N_2 \cdot U_2$

SAMPLE PROBLEM

Problem 25: A grocer wishes to mix one type of coffee worth $2.70 per pound with another type of coffee worth $1.95 per pound in order to make a 40 pound mixture worth $2.25 per pound. How many pounds of each type should he use?

PROCEDURE

(1) Represent the number of pounds of $2.70 coffee by P, and the number of pounds of $1.95 coffee by 40−P. (The total weight of the mixture is 40 pounds.)

(2) To obtain the value of each type of coffee, and the value of the mixture, multiply the number of pounds of each by the corresponding price per pound.

SOLUTION

(1) Let P = the number of pounds of $2.70 coffee,

and

40 − P = the number of pounds of $1.95 coffee.

(2) *Values*

$2.70 coffee: 2.70P

$1.95 coffee: 1.95(40−P)

mixture: 2.25(40)

(3) Write an equation expressing the fact that the value of the mixture is equal to the sum of the values of the two coffees. Solve this equation for P, and substitute the result into the expression for the number of pounds of $1.95 coffee.

(3) $\dfrac{\text{value of}}{\text{mixture}} = \dfrac{\text{value of}}{\$2.70 \text{ coffee}} + \dfrac{\text{value of}}{\$1.95 \text{ coffee}}$

$$2.25(40) = 2.70P + 1.95(40-P)$$
$$90 = 2.70P + 1.95(40) - 1.95(P)$$
$$90 = 2.70P + 78 - 1.95P$$
$$90 = .75P + 78$$
$$\underline{-78 = -78}$$
$$12 = .75P$$
$$\dfrac{12}{.75} = \dfrac{.75P}{.75}$$
$$16 = P$$

Thus, $40 - P = 24$

The grocer should use 16 pounds of the $2.70 coffee, and 24 pounds of the $1.95 coffee.   *Answer*

PRACTICE PROBLEMS

(1) What is the total value of a mixture of tea consisting of 20 pounds of tea worth 65 cents per pound and 10 pounds of tea worth 75 cents per pound?

(2) A 50 pound mixture of coffee consists of 35 pounds of coffee worth $4.00 per pound, and the rest of coffee worth D dollars per pound. What is the total value of the mixture, in dollars?

(3) Kathy wishes to mix candy worth 80 cents per pound with another type of candy worth 50 cents per pound in order to make 30 pounds of candy worth 75 cents per pound. How many pounds of the 80 cent candy should she use?

(4) Robin has one type of coffee worth $3.10 per pound, and another type of coffee worth $3.50 per pound. How many pounds of the $3.50 coffee should she use in order to make a 20 pound mixture of the two coffees worth $3.40 per pound?

(5) How many pounds of nuts worth 65 cents per pound must be mixed with 10 pounds of nuts worth 90 cents per pound in order to make a mixture worth 70 cents per pound?

Answers
(1) $20.50
(2) 140 + 15D
(3) 25 lbs.
(4) 15 lbs.
(5) 40 lbs.

12

12.1 MOTION PROBLEMS—BASIC FORMULAS

The distance traveled by a moving object is equal to its rate of speed multiplied by the time traveled. For example, an object traveling at the rate of 200 miles per hour (m.p.h.), for a period of 3 hours, will cover a distance of $(200) \cdot (3)$, or 600 miles. In this example, note that the rate can either mean a *constant* rate of 200 m.p.h., or an *average* rate of 200 m.p.h. It makes no difference.

In general, we have the following formula:

$$\boxed{\begin{array}{c} \text{Distance} = \text{Rate} \cdot \text{Time} \\ D = R \cdot T \end{array}}$$

By dividing both sides of this formula by R, we can change the subject of the formula to the time, T. Similarly, by dividing both sides of this formula by T, we can change the subject of the formula to the rate, R. Thus, we have the following alternate formulas:

$$D = R \cdot T \qquad\qquad D = R \cdot T$$
$$\frac{D}{R} = \frac{\cancel{R} \cdot T}{\cancel{R}} \qquad\qquad \frac{D}{T} = \frac{R \cdot \cancel{T}}{\cancel{T}}$$
$$\boxed{T = \frac{D}{R}} \qquad\qquad \boxed{R = \frac{D}{T}}$$

When using these formulas, remember that the units of measurement must match. For example, if the rate is measured in *miles per hour*, then the distance must be measured in *miles*, and the time must be measured in *hours*.

116 / Algebra Simplified and Self-Taught

SAMPLE PROBLEMS

Problem 26: A car traveled 78 miles in 98 minutes. If it traveled the first 42 miles in 50 minutes, what was its average rate of speed, in miles per hour, the rest of the way?

PROCEDURE

(1) Draw a diagram indicating the given information.

(2) To find the distance and time of the second part of the trip, subtract the distance and time of the first part of the trip from the total distance and time.

(3) Change the 48 minutes to hours by placing it over a denominator of 60 (60 minutes in 1 hour).

(4) Using the formula for the average rate, divide the distance of the second part of the trip, by the time of the second part of the trip.

SOLUTION

(1)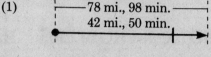

(2)

Distance	Time
78 miles	98 minutes
−42 miles	−50 minutes
36 miles	48 minutes

(3) $48 \text{ minutes} = \dfrac{48}{60} \text{ hours}$

$= \dfrac{4}{5} \text{ hours}$

(4) $R = \dfrac{D}{T}$

$R = \dfrac{36 \text{ miles}}{\dfrac{4}{5} \text{ hours}}$

$= 36 \div \dfrac{4}{5}$

$= \dfrac{\cancel{36}^{9}}{1} \times \dfrac{5}{\cancel{4}_{1}}$

$= 45 \text{ m.p.h.}$

Its average rate the rest of the way was 45 m.p.h.   *Answer*

Problem 27: A man runs a race in S seconds, running at an average rate of 14Y yards per second. If a woman runs the same race in $\frac{7}{8}$S seconds, what is her average rate, in yards per second?

PROCEDURE

(1) Draw a diagram indicating the given information.

(2) Represent the woman's rate by W.

SOLUTION

(1)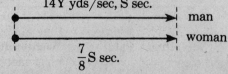

(2) Let W = the woman's rate.

PROCEDURE	SOLUTION
(3) Using the formula for the distance, multiply each runner's rate by their time.	(3) $D = R \cdot T$ $D_{man} = 14Y \cdot S = 14YS$ $D_{woman} = W \cdot \frac{7}{8}S = \frac{7}{8}WS$
(4) Since the distance of both runners is the same, set these expressions equal and solve for W.	(4) $D_{man} = D_{woman}$ $14YS = \frac{7}{8}WS$ $8 \cdot 14YS = \not{8} \cdot \frac{7}{\not{8}}WS$ $112YS = 7WS$ $\frac{112Y\not{S}}{7\not{S}} = \frac{\not{7}W\not{S}}{\not{7}\not{S}}$ $16Y = W$ The woman's rate is 16Y yds/sec. *Answer*

Problem 28: A train travels 120 miles at an average rate of 40 m.p.h. and returns along the same route at an average rate of 60 m.p.h. What is its average rate of speed for the entire trip?

PROCEDURE	SOLUTION
(1) Draw a diagram indicating the given information.	(1) out, 40 m.p.h., 120 mi., 60 m.p.h., back
(2) Using the formula for the time, find the time in each direction. Add the results.	(2) $T = \frac{D}{R}$ $T_{out} = \frac{120}{40} = 3$ hr. $T_{back} = \frac{120}{60} = 2$ hr. $T_{total} = T_{out} + T_{back}$ $= 3$ hr. $+ 2$ hr. $= 5$ hr.
(3) Using the formula for the average rate, divide the total distance (twice the one way distance) by the total time. (Note that the average rate we obtain is less than 50 m.p.h., the simple average of the two rates 40 m.p.h. and 60 m.p.h. The reason for this is that the train traveled more time at the slower rate, thus weighting the overall rate in that direction.)	(3) $R = \frac{D}{T}$ $= \frac{240 \text{ miles}}{5 \text{ hours}}$ $= 48$ m.p.h. The average rate for the entire trip is 48 m.p.h. *Answer*

118 / Algebra Simplified and Self-Taught

PRACTICE PROBLEMS

(1) From 10:30 A.M. to 11:45 A.M., Rick travels a total distance of 50 miles. What is his average rate of speed, in miles per hour?

(2) Michele can swim Y yards in S seconds. If in 20 seconds more time, Tim can swim twice as far as Michele does in S seconds, what is Tim's average rate, in yards per second?

(3) Deirdre runs a race in S seconds, running at an average rate of 10Y yards per second. If Maud finishes the race 3 seconds after Deirdre, what is Maud's average rate, in yards per second?

(4) Emmanuelle travels the first 60 miles of a trip at an average rate of 30 miles per hour. For the next 120 miles she averages 40 miles per hour. What is her average rate for the entire trip?

(5) On a trip of 408 miles, Dennis travels the first 160 miles in 4 hours. At what rate must he travel for the remainder of the trip in order to average 34 miles per hour for the entire trip?

Answers

(1) 40 m.p.h.
(2) $\dfrac{2Y}{S+20}$
(3) $\dfrac{10YS}{S+3}$
(4) 36 m.p.h.
(5) 31 m.p.h.

12.2 MOTION PROBLEMS—SPECIAL SITUATIONS

The situations described below are typical of many motion problems. In each of them, we can write an equation based on a relationship between certain distances. These relationships are derived from the accompanying diagrams.

SEPARATION SITUATION: Two objects start from the same place and travel in opposite directions. After a given amount of time, they are a certain distance apart.

$$\longleftarrow D_1 \quad\bullet\quad D_2 \longrightarrow$$
$$\longmapsto D_{apart} \longmapsto$$

The sum of the distances traveled by the two objects is equal to the final distance apart.

$$\boxed{D_1 + D_2 = D_{apart}}$$

MEETING SITUATION: Two objects start at a given distance apart and travel towards each other until they meet.

$$\bullet \longrightarrow D_1 \quad D_2 \longleftarrow \bullet$$
$$\longmapsto D_{apart} \longmapsto$$

The sum of the distances traveled by the two objects is equal to the original distance apart.

$$D_1 + D_2 = D_{apart}$$

OVERTAKE SITUATION: One object starts at a given place and travels in a certain direction. Some time later a second objects starts from the same place and travels at a faster rate in the same direction. The second object eventually overtakes the first object.

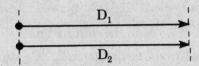

The distance traveled by the first object equals the distance traveled by the second object.

$$D_1 = D_2$$

ROUND TRIP SITUATION: One object travels out and back along the same path.

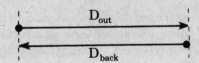

The distance out equals the distance back.

$$D_{out} = D_{back}$$

SAMPLE PROBLEMS

Problem 29: A freight train and a passenger train leave from the same station at 2:00 P.M. and travel in opposite directions. The freight train averages 52 m.p.h. and the passenger train averages 84 m.p.h. At what time are the trains 340 miles apart?

PROCEDURE	SOLUTION
(1) Represent the number of hours traveled by each train by h. (They both travel the same amount of time.)	(1) Let $h =$ the number of hours traveled by each train.
(2) Using the distance formula, multiply each rate by h.	(2) $D = R \cdot T$ $D_{freight} = 52h$ $D_{passenger} = 84h$

120 / *Algebra Simplified and Self-Taught*

(3) Draw a diagram showing all distances.

(3)
```
       freight        passenger
        52h              84h
    |----------•------------|
    |--------- 340 mi. -----|
```

(4) Write an equation expressing the fact that the sum of the distances traveled by the two trains equals the final distance apart. Solve this equation for h.

(4)
$$52h + 84h = 340$$
$$136h = 340$$
$$\frac{\cancel{136}h}{\cancel{136}} = \frac{340}{136}$$
$$h = 2\tfrac{1}{2} \text{ hours}$$

(5) Add the time traveled to the starting time, 2:00 P.M.

(5) 2:00 P.M. $+ 2\tfrac{1}{2}$ hrs. $= 4{:}30$ P.M.

The trains are 340 mi. apart at 4:30 P.M. *Answer*

Problem 30: Two cars start traveling towards each other at the same time from cities 416 miles apart. The average rate of one car is 10 m.p.h. faster than the other car. If they pass each other in 4 hours, what is the average rate of each car?

| PROCEDURE | SOLUTION |

(1) Represent the rate of the slower car by s and the rate of the faster car by $s+10$ ("10 m.p.h. faster").

(1) Let $s=$ the rate of the slower car,
and $s+10=$ the rate of the faster car.

(2) Using the distance formula, multiply each rate by 4 hrs. (Since both cars leave at the same time, they both travel the same amount of time.)

(2)
$$D = R \cdot T$$
$$D_{\text{slower}} = s \cdot 4 = 4s$$
$$D_{\text{faster}} = (s+10) \cdot 4 = 4s + 40$$

(3) Draw a diagram showing all distances.

(3)
```
    slower          faster
     4s              4s+40
  •------->|<-----------•
  |------- 416 mi. ------|
```

(4) Write an equation expressing the fact that the sum of the distances traveled by the two cars equals the original distance apart. Solve this equation for s, and substitute the result into the expression for the rate of the faster car.

(4)
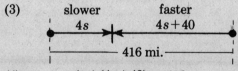
$$s = 47$$
Thus, $s + 10 = 57$

The slower car averages 47 m.p.h., and the faster car averages 57 m.p.h. *Answer*

Problem 31: A man leaves his home and starts jogging at an average rate of 8 m.p.h. Fifteen minutes later his wife leaves home and starts bicycling after him at an average rate of 11 m.p.h. How long after the wife leaves does she catch up with her husband?

	PROCEDURE		SOLUTION
(1)	Represent the time of the wife (in hours) by h, and the time of the husband by $h+\frac{1}{4}$. (The husband leaves 15 min., or $\frac{1}{4}$ hour, earlier than his wife.)	(1)	Let $h=$ the number of hours traveled by the wife, and $h+\frac{1}{4}=$ the number of hours traveled by the husband.
(2)	Using the distance formula, multiply each rate by its respective time.	(2)	$D = R \cdot T$ $D_{\text{wife}} = 11 \cdot h = 11h$ $D_{\text{husband}} = 8 \cdot (h+\frac{1}{4}) = 8h+2$
(3)	Draw a diagram showing all distances.	(3)	$\overset{11h}{\longrightarrow}$ wife $\overset{8h+2}{\longrightarrow}$ husband
(4)	Write an equation expressing the fact that the distance traveled by the wife and the husband is the same. Solve this equation for h.	(4)	$11h = 8h+2$ $-8h = -8h$ $\overline{}$ $3h = 2$ $\frac{\cancel{3}h}{\cancel{3}} = \frac{2}{3}$ $h = \frac{2}{3}$ hr., or 40 min.

The wife catches up with the husband in 40 min.   *Answer*

Problem 32: A plane makes a round trip flight from the San Francisco airport which lasts 5 hours. If the average rate out is 200 m.p.h. and the average rate back is 300 m.p.h., what is the plane's greatest distance from the airport?

	PROCEDURE		SOLUTION
(1)	Represent the time flying out by h, and the time flying back by $5-h$. (The total flight time is 5 hours.)	(1)	Let $h=$ the number of hours flying out, and $5-h=$ the number of hours flying back.
(2)	Using the distance formula, multiply each rate by its respective time.	(2)	$D = R \cdot T$ $D_{\text{out}} = 200h$ $D_{\text{back}} = 300(5-h)$
(3)	Draw a diagram showing all distances.	(3)	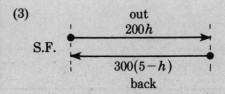

(4)	Write an equation expressing the fact that the distance out equals the distance back. Solve this equation for h.	(4)	$200h = 300(5-h)$ $200h = 1500 - 300h$ $\underline{+300h = +300h}$ $500h = 1500$ $\dfrac{500h}{500} = \dfrac{1500}{500}$ $h = 3 \text{ hours}$
(5)	Substitute this value into the expression for the distance out.	(5)	$D_{out} = 200h$ $D_{out} = 200(3)$ $\phantom{D_{out}} = 600 \text{ miles}$

The plane's greatest distance from the airport is 600 mi.   *Answer*

PRACTICE PROBLEMS

(1) Two trains start at the same time from the same station and travel in opposite directions. One travels at the rate of 50 m.p.h., and the other travels at the rate of 60 m.p.h. In how many hours are the trains 440 miles apart?

(2) Maria took a trip of 370 miles by bus and train. She traveled 3 hours by bus and 4 hours by train. If the train averaged 19 m.p.h. more than the bus, what was the rate of the train?

(3) At 5 P.M., Jill leaves New York City and travels toward Boston at an average rate of 56 m.p.h. At 6 P.M., Billy leaves Boston and travels toward New York, on the same road as Jill, at an average rate of 50 m.p.h. If they were initially 215 miles apart, at what time do Billy and Jill pass each other?

(4) An airplane leaves LaGuardia Airport at 3 P.M. and flies west at an average rate of 200 m.p.h. At 3:30 P.M., another airplane leaves the airport and flies west at an average rate of 250 m.p.h. At what time does the second plane overtake the first plane?

(5) A helicopter makes a round trip flight which lasts $4\frac{1}{2}$ hours. If the average rate out was 80 m.p.h., and the average rate back was 100 m.p.h., what was the helicopter's greatest distance from the airport?

Answers
(1) 4 hours
(2) 61 m.p.h.
(3) 7:30 P.M.
(4) 5:30 P.M.
(5) 200 miles

13

13.1 WORK PROBLEMS—INDIVIDUALS

In work problems involving individuals (individual people, individual machines, etc.), we are told how long it takes several individuals working alone to complete a job, and are asked to determine how long it would take the same individuals to complete the job, working together.

If an individual can complete a job in 5 hours, then after 1 hour he will complete $\frac{1}{5}$ of the job, after 2 hours he will complete $\frac{2}{5}$ of the job, after 3 hours he will complete $\frac{3}{5}$ of the job, and so on. (This assumes, of course, that the individual works at a constant rate.) In general:

> The Fractional Part of the Job Completed =
> $$\frac{\text{Actual Time Worked}}{\text{Total Time Required}}$$

When individuals work together, the fractional part of the job they complete as a group is simply the *sum* of the fractional parts they complete as individuals within the group (assuming that they work independently of each other). For example, if one individual can complete a job in 6 hours alone, and another individual can complete the same job in 12 hours alone, then after working together for 1 hour they complete $\frac{1}{6} + \frac{1}{12}$, or $\frac{1}{4}$ of the job, after working together for 2 hours they complete $\frac{2}{6} + \frac{2}{12}$, or $\frac{1}{2}$ of the job, and so on. In order for them to complete the job, the sum of the fractional parts must equal 1 (1 whole job). Thus, if we let T represent the amount of time required to complete the job together, we get the equation

$$\frac{T}{6} + \frac{T}{12} = 1$$

To solve this equation for T, we multiply both sides by the common denominator 12, and get

$$12\left(\frac{T}{6} + \frac{T}{12}\right) = 12(1)$$

$$2T + T = 12$$

$$3T = 12$$

$$T = 4 \text{ hours}$$

In other words, the two individuals would complete the job in 4 hours working together.

In general,

> If one individual can complete a job in H_1 hours alone, and another individual can complete the same job in H_2 hours alone, then the amount of hours it would take them to complete the job working together, T, is given by the equation
>
> $$\frac{T}{H_1} + \frac{T}{H_2} = 1$$

SAMPLE PROBLEMS

Problem 33: If used alone, one pipe can fill a tank in 8 minutes, a second pipe can fill the tank in 12 minutes, and a third pipe can fill the tank in 24 minutes. How long would it take all three pipes, operating together, to fill the tank?

PROCEDURE	SOLUTION
(1) Represent the number of minutes required to fill the tank by M.	(1) Let M = the number of minutes required by the three pipes to fill the tank together.
(2) For each pipe, write a fraction indicating the part of the job it completes in M minutes. (Each pipe operates for M minutes.)	(2) 8 min. pipe: $\frac{M}{8}$ 12 min. pipe: $\frac{M}{12}$ 24 min. pipe: $\frac{M}{24}$
(3) Write an equation expressing the fact that, since the job is completed, the sum of the fractional parts must equal 1. Solve this equation for M.	(3) $\frac{M}{8} + \frac{M}{12} + \frac{M}{24} = 1$ $24\left(\frac{M}{8} + \frac{M}{12} + \frac{M}{24}\right) = 24(1)$ $3M + 2M + M = 24$ $6M = 24$ $\frac{6M}{6} = \frac{24}{6}$ $M = 4$ min. The three pipes, operating together, will fill the tank in 4 minutes. *Answer*

Problem 34: Operating together, two copy machines can reproduce a report in 30 minutes. If one machine is twice as fast as the other, how long would it take each machine, operating alone, to do the same job?

PROCEDURE	SOLUTION
(1) Represent the time of the faster machine by T, and the time of the slower machine by 2T. (The slower machine requires twice as much time.)	(1) Let T = the time required by the faster machine to complete the job alone, and 2T = the time required by the slower machine to complete the job alone.
(2) For each machine, write a fraction indicating the part of the job completed in 30 minutes.	(2) Faster Machine: $\dfrac{30}{T}$ Slower Machine: $\dfrac{30}{2T}$
(3) Write an equation expressing the fact that when the job is completed, the sum of the fractions must be 1. Solve this equation for T, and substitute the result into the expression for the time of the slower machine.	(3) $\dfrac{30}{T} + \dfrac{30}{2T} = 1$ $2T\left(\dfrac{30}{T} + \dfrac{30}{2T}\right) = 2T(1)$ $2T\left(\dfrac{30}{T}\right) + 2T\left(\dfrac{30}{2T}\right) = 2T$ $60 + 30 = 2T$ $90 = 2T$ $\dfrac{90}{2} = \dfrac{2T}{2}$ $45 = T$ Thus, 2T = 90 The faster machine would take 45 min. operating alone, and the slower machine would take 90 min. operating alone. *Answer*

PRACTICE PROBLEMS

(1) Working alone, Danny can paint a room in 4 hours and Cindy can paint the same room in 5 hours. After working together for 2 hours, what fraction of the room remains to be painted?

(2) Brita can rake a lawn in 45 minutes by herself, and Denise can rake the same lawn in 90 minutes by herself. How long would it take the two of them working together to rake the lawn?

Answers
(1) $\dfrac{1}{10}$
(2) 30 min.

(3) Working alone, Fritz can build a bookcase in 8 hours and Holly can build the same bookcase in 10 hours. After working together for 3 hours, Fritz had to leave. How long will it take Holly to complete the bookcase by herself?

(4) Working together, Nora, Weedy, and Daphne can set the Thanksgiving table in 4 minutes. If Nora, alone, can set the table in 7 minutes, and Weedy, alone, can set the table in 28 minutes, how long would it take Daphne to set the table by herself?

(5) One printing press is three times faster than another. If together the two presses can print a newspaper in 6 hours, how long would it take the faster press to print the newspaper by itself?

> **Answers**
> (3) $3\frac{1}{4}$ hrs.
> (4) 14 min.
> (5) 8 hrs.

13.2 WORK PROBLEMS—GROUPS

In work problems involving groups, we are told how long it takes a group of a certain size to complete a job, and are asked to determine how long it would take a group of a different size to complete the same job. For example:

Suppose a group of 4 machines requires 6 hours to complete a job. If the number of machines is doubled to 8, the number of hours would be cut in half to 3. If, on the other hand, the number of machines is cut in half to 2, the number of hours would be doubled to 12. In other words, the more machines, the less time; the less machines, the more time.

In each case, the product of the number of machines and the number of hours is the same, 24. That is, each combination must produce the same amount of work to complete the job, 24 "machine-hours."

In general,

# of machines		# of hours		
4	×	6	=	24
8	×	3	=	24
2	×	12	=	24

> If a group of size N_1 completes a job in H_1 hours, and another group of size N_2 completes the same job in H_2 hours, then
>
> $$N_1 \cdot H_1 = N_2 \cdot H_2$$

Note that this principle assumes that the individual workers in one group work at the same rate as the individual workers in the other group.

SAMPLE PROBLEM

Problem 35: Eight machines can complete a job in 5 hours. If only 6 of these machines are working, how long will it take them to complete the same job?

PROCEDURE	SOLUTION
(1) Represent the number of hours required by the smaller group by h.	(1) Let h = the number of hours required by the group of 6 machines to complete the job.
(2) Write an equation expressing the fact that the product of the number of machines and the number of hours must be the same for both groups. Solve this equation for h.	(2) $N_1 \cdot H_1 = N_2 \cdot H_2$ $$8 \cdot 5 = 6 \cdot h$$ $$40 = 6h$$ $$\frac{40}{6} = \frac{6h}{6}$$ $$6\tfrac{2}{3} = h$$ The group of 6 machines will take $6\tfrac{2}{3}$ hours to complete the job. *Answer*

PRACTICE PROBLEMS

Answers
(1) 10 hrs.
(2) $\dfrac{QD}{D-2}$
(3) $\dfrac{MH}{M-N}$
(4) 4 days
(5) 2 pipes

(1) Five tractors can plow a certain field in 12 hours. How long would it take 6 of these tractors to plow the same field?

(2) If Q machines can complete a job in D days, how many of these machines are required to complete the same job in 2 days less time?

(3) M machines can complete a job in H hours. If N machines break down before starting the job, how many hours will it take the remaining number of machines to do the same job?

(4) Four people can build a fence in 12 days. How much less time would be required by 6 people working at the same rate?

(5) If 4 pipes can fill a pool in 15 hours, how many more of these pipes are required to fill the same pool in only 10 hours?

14

14.1 SET PROBLEMS

A collection of things (numbers, people, attributes, etc.) is called a **set**. Each member of a set is called an **element** of the set. For example, the set of odd numbers less than 10 consists of the elements 1, 3, 5, 7, and 9.

A common way of depicting the relationship between sets is by a diagram, called a **Venn diagram**, in which each set is represented by a circle. All the possible Venn diagrams for two sets are shown below. In the diagram on the left, set A and set B have *no elements* in common and are said to be **disjoint**; in the next diagram, set A and set B have *some elements* in common (but not all) and are said to be **overlapping**; in the next diagram, set A has *all its elements* in set B and is said to be a **subset** of set B; in the last diagram, set B has all its elements in set A and is thus a subset of set A.

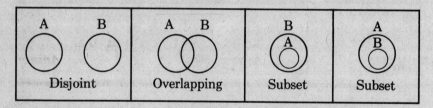

Venn diagrams are particularly useful in solving problems involving *overlapping* sets. For example, suppose we are told that in a class of 25 students, 15 students have brown hair, 12 students have blue eyes, and 8 students have *both* brown hair *and* blue eyes. This information can be organized into a Venn diagram, as shown below. The overlapping section contains the 8 students who have both brown hair and blue eyes. This leaves 7 students (15−8=7) in the section of brown hair, but not blue eyes, and 4 students (12−8=4) in the section of blue eyes, but not brown hair.

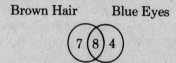

By organizing the information this way, we can then determine, for example, how many students have *neither* brown hair *nor* blue eyes. This is simply the number of students outside the two circles. Since there are 19 students inside the two circles (7+8+4=19) and 25 students in the entire class, we can conclude that there are 6 students (25−19=6) who have neither brown hair nor blue eyes.

Remember that, in general, it is best to fill in the number of elements in the overlapping section first. This will help avoid the common error of counting the same element more than once.

SAMPLE PROBLEM

Problem 36: In a survey, 60% of the people said that they had tried brand X, 50% said that they had tried brand Y, and 20% said that they had tried both. What percent said that they had tried neither?

PROCEDURE

(1) Draw a Venn diagram depicting brand X and brand Y.

(2) Place the 20% that had tried both brands in the overlapping section. This leaves 40% in the brand X, but not brand Y section, and 30% in the brand Y, but not brand X section.

(3) To obtain the percent outside the two circles (the percent that had tried neither), subtract the sum of the percents inside the two circles from 100%.

SOLUTION

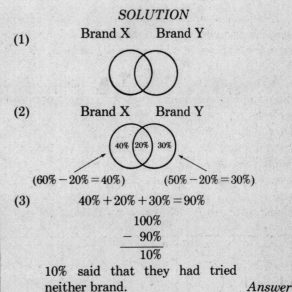

(3) 40% + 20% + 30% = 90%

$$\begin{array}{r} 100\% \\ -\ 90\% \\ \hline 10\% \end{array}$$

10% said that they had tried neither brand.   *Answer*

PRACTICE PROBLEMS

Answers
(1) 8
(2) 20%
(3) 15
(4) 27

(1) In a group of 30 students, 18 are taking calculus, 15 are taking physics, and 11 are taking both calculus and physics. How many students in the group are taking neither calculus nor physics?

(2) In a recent survey, 40% of the people said they read newspaper X, 50% said they read newspaper Y, and 10% said they read both newspapers. What percent of the people said they read neither newspaper?

(3) At a party of 100 people, 75 people wore bluejeans, 30 people wore sweaters, and 20 people wore both. How many people at the party wore neither bluejeans nor sweaters?

(4) At a wedding reception of 50 people, 37 people had an appetizer and 30 people had a dessert. If 10 people had neither an appetizer nor a dessert, how many people at the reception had both?

(5) In a foreign language school, 64% of the students are studying Spanish, 52% are studying French, and 28% are studying a language other than Spanish or French. What percent of the students at the school are studying both Spanish and French?

Answer
(5) 44%

WORD PROBLEMS PRACTICE TEST

1. At a party there are S men and T women. If 3 more men arrive, and 2 women leave, what fraction of the party are now women?

   (A) $\dfrac{T-2}{S-T+5}$

   (B) $\dfrac{T-S}{S+T}$

   (C) $\dfrac{T-2}{S+T+1}$

   (D) $\dfrac{T}{S+T+1}$

   (E) $\dfrac{T-2}{S+3}$

2. A traveler spends $\frac{2}{5}$ of his money on a plane ticket and $\frac{1}{3}$ of his money on lodging. If he is left with $160, how much money did he start with?

   (A) $600
   (B) $200
   (C) $1,200
   (D) $218
   (E) $540

3. On a driving test, 4 out of 5 applicants passed the written test. Of every 4 that passed the written test, 1 failed the road test. What fraction of all the applicants passed both tests?

   (A) $\dfrac{3}{4}$

   (B) $\dfrac{4}{15}$

   (C) $\dfrac{5}{8}$

   (D) $\dfrac{3}{5}$

   (E) $\dfrac{8}{15}$

4. If M calculators can be bought for D dollars, how many calculators can be bought for G dollars, at the same rate?

   (A) $\dfrac{M}{DG}$

   (B) $M(D-G)$

   (C) $\dfrac{MG}{D}$

   (D) $\dfrac{MD}{G}$

   (E) $\dfrac{DG}{M}$

5. In a school debating team, there are three times as many juniors as sophomores, and twice as many seniors as juniors. What percent of the debating team are juniors?

   (A) 60%
   (B) 30%
   (C) 10%
   (D) 22.2%
   (E) 16%

6. In a recent election, only three candidates, A, B, and C, received votes. If A received $\frac{1}{5}$ of the votes, B received 35% of the votes, and C received 900 votes, how many votes did A receive?

   (A) 700
   (B) 2,000
   (C) 782
   (D) 400
   (E) 1,100

7. The price of an airline ticket increases from $240 to $300. What is the percent of increase?

   (A) 80%
   (B) 25%

(C) 30%
(D) 120%
(E) 20%

8. Nikki sells her house for $50,000, thus making a profit of 25% of the cost. What did the house cost?

(A) $40,000
(B) $37,500
(C) $30,000
(D) $42,500
(E) $20,000

9. If a selling price of $24 results in a 20% discount off the list price, what selling price would result in a 30% discount off the list price?

(A) $9
(B) $20
(C) $27
(D) $18
(E) $21

10. A markup of 10%, followed by a discount of 30%, is equivalent to a single discount of what percent?

(A) 28%
(B) 20%
(C) 23%
(D) 25%
(E) 37%

11. Four times the sum of a number and 2 is equal to 11 more than the number. What is the number?

(A) 1
(B) $4\frac{1}{3}$
(C) 3
(D) 2
(E) $1\frac{4}{5}$

12. Joe has 20 coins consisting only of dimes and quarters. If the number of dimes is 5 more than twice the number of quarters, what is the total value of the coins?

(A) $3.20
(B) $3.75
(C) $2.75
(D) $4.25
(E) $4.50

13. If the sum of three consecutive even integers is twice the largest, what is the largest?

(A) 8
(B) 6
(C) 7
(D) 2
(E) 10

14. Ann is now twice as old as Jerry. Four years ago she was three times as old as Jerry was then. How old is Ann now?

(A) 4
(B) 16
(C) 24
(D) 8
(E) 10

15. A basketball team averages 98 points in its first three games. How many points must it score in the next game in order to have a 100 point average overall?

(A) 102
(B) 104
(C) 108
(D) 112
(E) 106

16. In a certain company, 5 employees earn $16,000 a year, 3 employees earn $20,000 a year, and 2 employees earn $24,000 a year. What is the average annual salary of these employees?

(A) $18,800
(B) $20,000
(C) $19,729
(D) $19,200
(E) $18,000

17. Joan wishes to mix some tea worth $1.40 per pound with another type of tea worth $2.25 per pound in order to make a 10 pound mixture worth $1.74 per pound. How many pounds of the $1.40 tea should she use?

(A) 4
(B) 5
(C) 6
(D) 7
(E) 8

18. Maureen runs a 26 mile marathon in 5.4 hours. If she runs the first half of the race in 2.8 hours, what is her average rate, in miles per hour, for the second half of the race?

   (A) 2
   (B) 2.6
   (C) 6
   (D) 5
   (E) 4.8

19. Rick travels 120 miles at an average rate of 20 m.p.h., and returns along the same route at an average rate of 30 m.p.h. What is his average rate, in miles per hour, for the entire round trip?

   (A) 26
   (B) 25
   (C) 22
   (D) 24
   (E) 23

20. A train leaves a station at 6 P.M. and travels east at an average rate of 80 m.p.h. At 8 P.M. another train leaves the same station and travels west at an average rate of 90 m.p.h. At what time are the two trains 415 miles apart?

   (A) 9:38 P.M.
   (B) 10:44 P.M.
   (C) 11:30 P.M.
   (D) 10:30 P.M.
   (E) 9:30 P.M.

21. Tom leaves a roadside diner and travels south at an average rate of 45 m.p.h. One hour later, Rosemary leaves the same diner and travels south along the same road at an average rate of 60 m.p.h. How many miles are the two people from the diner when Rosemary catches up with Tom?

   (A) 180
   (B) 120
   (C) 135
   (D) 360
   (E) 240

22. Working alone, Allen can build a fence in 8 hours, and Danny can build the same fence in 16 hours. If after working alone for 2 hours, Allen is joined by Danny, how many more hours will it take the two of them to complete the fence?

   (A) 2
   (B) $3\frac{1}{3}$
   (C) 6
   (D) 10
   (E) 4

23. One pipe can fill a tank three times as fast as another pipe. If together the two pipes can fill the tank in 36 minutes, how many minutes would it take the slower pipe to fill the tank by itself?

   (A) 192
   (B) 108
   (C) 81
   (D) 144
   (E) 120

24. Five printing presses can print a newspaper in 6 hours. If only 3 of the presses are operating, how long will it take them to do the same job?

   (A) 8
   (B) 9
   (C) 10
   (D) 11
   (E) 12

25. In a group of 15 married couples, 16 people have brown hair, 12 people have blue eyes, and 9 people have both brown hair and blue eyes. How many people have neither brown hair nor blue eyes?

   (A) 2
   (B) 11
   (C) 14
   (D) 16
   (E) 19

WORD PROBLEMS PRACTICE TEST—ANSWER KEY

1. C	6. D	11. A	16. A	21. A
2. A	7. B	12. C	17. C	22. E
3. D	8. A	13. B	18. D	23. D
4. C	9. E	14. B	19. D	24. C
5. B	10. C	15. E	20. E	25. B

WORD PROBLEMS PRACTICE TEST—ANSWERS AND SOLUTIONS

1.(C) Men $= S+3$ (3 men arrive)
  Women $= T-2$ (2 women leave)
  Total $= S+T-1$

Fraction of women $= \dfrac{\text{Women}}{\text{Total}}$
$= \dfrac{T-2}{S+T+1}$

2.(A) The fraction of money spent $= \dfrac{2}{5} + \dfrac{1}{3} = \dfrac{11}{15}$
Thus, the fraction of money left is
$1 - \dfrac{11}{15} = \dfrac{4}{15}$
Let $S =$ the amount of money started with.
Thus,
$$\dfrac{4}{15} S = \$160$$
$$15 \cdot \dfrac{4}{15} S = 15 \cdot 160$$
$$4S = 2400$$
$$S = \$600$$

3.(D) $\dfrac{4}{5}$ passed the written test.

$\dfrac{3}{4}$ of those that passed the written test also passed the road test.

Thus, the fraction that passed both tests is
$\dfrac{3}{4}$ of $\dfrac{4}{5} = \dfrac{3}{4} \times \dfrac{4}{5} = \dfrac{3}{5}$

4.(C) Let $x =$ the number of calculators that can be bought for G dollars.

$\dfrac{\text{calculators}}{\text{dollars}} \rightarrow \dfrac{M}{D} = \dfrac{x}{G}$ (cross-multiply)
$Dx = MG$
$x = \dfrac{MG}{D}$

5.(B) Let So $=$ the number of sophomores,
  Jr $=$ the number of juniors,
  and Se $=$ the number of seniors.

Jr $= 3$So   or   So $= \dfrac{\text{Jr}}{3}$
Se $= 2$Jr

Fraction of juniors $= \dfrac{\text{Jr}}{\text{So} + \text{Jr} + \text{Se}}$

(substitute So $= \dfrac{\text{Jr}}{3}$ and Se $= 2$Jr)

$= \dfrac{\text{Jr}}{\dfrac{\text{Jr}}{3} + \text{Jr} + 2\text{Jr}}$ (multiply by 3)

$= \dfrac{3 \cdot \text{Jr}}{3 \cdot \dfrac{\text{Jr}}{3} + 3 \cdot \text{Jr} + 3 \cdot 2\text{Jr}}$

$= \dfrac{3\text{Jr}}{\text{Jr} + 3\text{Jr} + 6\text{Jr}}$

$= \dfrac{3\text{Jr}}{10\text{Jr}}$

$= \dfrac{3}{10}$

Thus, the percent of juniors $= \dfrac{3}{10} \times 100\% = 30\%$

6.(D) A received $\dfrac{1}{5}$, or 20%, of the vote.
B received 35% of the vote.
Thus, C received $100\% - (20\% + 35\%) = 45\%$ of the vote.

Let T $=$ the total number of votes cast.
Since C received 900 votes,

$$\frac{\text{Part}}{\text{Whole}} = \frac{\text{Percent}}{100\%}$$
$$\frac{900}{T} = \frac{45}{100} \text{ (cross-multiply)}$$
$$45T = 90000$$
$$T = 2000 \text{ total votes}$$

Thus, A received $\frac{1}{5}$ of 2000

$= \frac{1}{5} \times 2000 = 400$ votes.

7.(B) $\frac{\text{amount of increase}}{\text{original value}} = \frac{\% \text{ of increase}}{100\%}$
$$\frac{\$300 - \$240}{\$240} = \frac{x}{100}$$
$$\frac{60}{240} = \frac{x}{100} \text{ (cross-multiply)}$$
$$240x = 6000$$
$$x = 25\% \text{ increase}$$

8.(A) Let C = the cost of the house.

Selling Price = (100% + % of Profit) × Cost
$$\$50{,}000 = (100\% + 25\%) \times C$$
$$50{,}000 = 1.25C$$
$$\frac{50{,}000}{1.25} = \frac{1.25C}{1.25}$$
$$\$40{,}000 = C$$

9.(E) Let L = the list price for a 20% discount.

Selling Price = (100% − % of Discount) × List Price
$$\$24 = (100\% - 20\%) \times L$$
$$24 = .80L$$
$$\frac{24}{.80} = \frac{.80L}{.80}$$
$$\$30 = L$$

Let S = the selling price for a 30% discount.

Selling Price = (100% − % of Discount) × List Price
$$S = (100\% - 30\%) \times \$30$$
$$= .70 \times 30$$
$$= \$21$$

10.(C) After a 10% markup, the selling price is 110% of the original value.
After a 30% discount, the selling price is 70% of the original value.
Thus, a 10% markup followed by a 30% discount means
$$.70 \times 1.10 \times \text{original value}$$
$$= .77 \times \text{original value}$$
This is equivalent to a single discount of 23% off the original value.

11.(A) Let n = the number.
$$4(n+2) = n+11$$
$$4n+8 = n+11$$
$$3n+8 = 11$$
$$3n = 3$$
$$n = 1$$

12.(C) Let Q = the number of quarters,
and 2Q + 5 = the number of dimes.

Since there are a total of 20 coins,
$$Q + 2Q + 5 = 20$$
$$3Q + 5 = 20$$
$$3Q = 15$$
$$Q = 5 \text{ quarters}$$
and $2Q + 5 = 2(5) + 5$
$$= 15 \text{ dimes}$$
Thus, the total value = 5($.25) + 15($.10) = $2.75

13.(B) Let N = the first even integer,
N + 2 = the second even integer,
and N + 4 = the third even integer.
$$N + N + 2 + N + 4 = 2(N+4)$$
$$3N + 6 = 2N + 8$$
$$N + 6 = 8$$
$$N = 2$$
Thus, the largest, N + 4 = 2 + 4 = 6

14.(B) Let J = Jerry's age now,
and 2J = Ann's age now.
Thus, J − 4 = Jerry's age 4 years ago,
and 2J − 4 = Ann's age 4 years ago.
$$2J - 4 = 3(J - 4)$$
$$2J - 4 = 3J - 12$$
$$-4 = J - 12$$
$$8 = J$$

Thus, Ann's age now = 2J = 2(8) = 16

15.(E) Let S = the score of the next game.
$$\text{Average} = \frac{\text{Sum}}{N}$$
$$100 = \frac{3(98) + S}{4}$$
$$100 = \frac{294 + S}{4}$$
$$400 = 294 + S$$
$$106 = S$$

16.(A) Combined Average =
$$\frac{5(\$16{,}000) + 3(\$20{,}000) + 2(\$24{,}000)}{5+3+2}$$
$$= \frac{\$80{,}000 + \$60{,}000 + \$48{,}000}{10}$$
$$= \$18{,}800$$

17.(C) Let P = the number of pounds of $1.40 tea,
and 10 − P = the number of pounds of $2.25 tea.

Value of $1.40 tea = $1.40P
Value of $2.25 tea = $2.25(10 − P)
Value of mixture = $1.74(10)
$$1.74(10) = 1.40P + 2.25(10 - P)$$
$$17.4 = 1.40P + 22.5 - 2.25P$$
$$17.4 = 22.5 - .85P$$
$$-5.1 = -.85P$$
$$6 = P$$

18.(D)

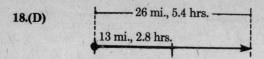

The distance in the second half =
26 mi. − 13 mi. = 13 mi.
The time in the second half =
5.4 hrs. − 2.8 hrs. = 2.6 hrs.

$$\text{Rate} = \frac{\text{Distance}}{\text{Time}}$$

$$R = \frac{13 \text{ mi.}}{2.6 \text{ hrs.}}$$

= 5 mi./hr. in second half

19.(D)

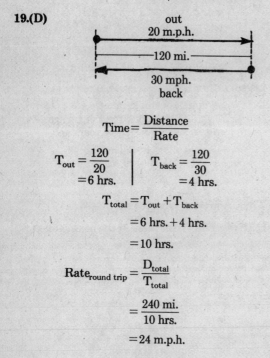

$$\text{Time} = \frac{\text{Distance}}{\text{Rate}}$$

$T_{out} = \dfrac{120}{20}$ $\Big|$ $T_{back} = \dfrac{120}{30}$
= 6 hrs. = 4 hrs.

$T_{total} = T_{out} + T_{back}$
= 6 hrs. + 4 hrs.
= 10 hrs.

$$\text{Rate}_{\text{round trip}} = \frac{D_{total}}{T_{total}}$$

$$= \frac{240 \text{ mi.}}{10 \text{ hrs.}}$$

= 24 m.p.h.

20.(E) Let h = the number of hours traveled by the train which leaves at 6 P.M., and $h-2$ = the number of hours traveled by the train which leaves at 8 P.M.

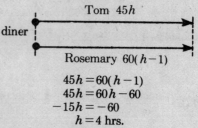

$90(h-2) + 80h = 415$
$90h - 180 + 80h = 415$
$170h - 180 = 415$
$170h = 595$
$h = 3.5$ hrs

Thus, the trains are 415 miles apart at 6 P.M. + 3.5 hrs = 9:30 P.M.

21.(A) Let h = the number of hours traveled by Tom, and $h-1$ = the number of hours traveled by Rosemary.

diner — Tom $45h$
Rosemary $60(h-1)$

$45h = 60(h-1)$
$45h = 60h - 60$
$-15h = -60$
$h = 4$ hrs.

Thus, the number of miles from the diner
$= 45h = 45(4) = 180$ mi.

22.(E) Let h = the number of hours worked together to complete the fence.

Allen alone   Allen   Danny
$\dfrac{2}{8} + \dfrac{h}{8} + \dfrac{h}{16} = 1$ (multiply by 16)

$\cancel{16} \cdot \dfrac{2}{\cancel{8}} + \cancel{16} \cdot \dfrac{h}{\cancel{8}} + \cancel{16} \cdot \dfrac{h}{\cancel{16}} = 16 \cdot 1$

$4 + 2h + h = 16$
$4 + 3h = 16$
$3h = 12$
$h = 4$ hrs.

23.(D) Let M = the number of minutes required by the faster pipe to fill the tank alone, and 3M = the number of minutes required by the slower pipe to fill the tank alone.

$\dfrac{36}{M} + \dfrac{36}{3M} = 1$ (multiply by 3M)

$3M \cdot \dfrac{36}{M} + 3M \cdot \dfrac{36}{3M} = 3M \cdot 1$

$108 + 36 = 3M$
$144 = 3M$
48 min. $= M$

Thus, the number of minutes required by the slower pipe $= 3M = 3(48) = 144$ min.

24.(C) Let h = the number of hours required by the group of 3 presses to print the newspaper.

5 presses · 6 hours = 3 presses · h
$30 = 3h$
10 hrs. $= h$

25.(B)

brown hair   blue eyes
7 9 3

$(16 - 9 = 7)$ $(12 - 9 = 3)$

Since there are a total of 30 people (15 couples), the number of people that have neither brown hair nor blue eyes is $30 - (7 + 9 + 3) = 30 - 19 = 11$.